Tim Staufenberger

Chitinases in the tree of life

Tim Staufenberger

Chitinases in the tree of life

Ecological, kinetic and structural studies of
archaeal and marine bacterial chitinases

Südwestdeutscher Verlag für Hochschulschriften

Impressum/Imprint (nur für Deutschland/only for Germany)
Bibliografische Information der Deutschen Nationalbibliothek: Die Deutsche Nationalbibliothek verzeichnet diese Publikation in der Deutschen Nationalbibliografie; detaillierte bibliografische Daten sind im Internet über http://dnb.d-nb.de abrufbar.
Alle in diesem Buch genannten Marken und Produktnamen unterliegen warenzeichen-, marken- oder patentrechtlichem Schutz bzw. sind Warenzeichen oder eingetragene Warenzeichen der jeweiligen Inhaber. Die Wiedergabe von Marken, Produktnamen, Gebrauchsnamen, Handelsnamen, Warenbezeichnungen u.s.w. in diesem Werk berechtigt auch ohne besondere Kennzeichnung nicht zu der Annahme, dass solche Namen im Sinne der Warenzeichen- und Markenschutzgesetzgebung als frei zu betrachten wären und daher von jedermann benutzt werden dürften.

Coverbild: www.ingimage.com

Verlag: Südwestdeutscher Verlag für Hochschulschriften GmbH & Co. KG
Heinrich-Böcking-Str. 6-8, 66121 Saarbrücken, Deutschland
Telefon +49 681 37 20 271-1, Telefax +49 681 37 20 271-0
Email: info@svh-verlag.de

Approved by: Kiel, CAU, Diss., 2012

Herstellung in Deutschland (siehe letzte Seite)
ISBN: 978-3-8381-3341-6

Imprint (only for USA, GB)
Bibliographic information published by the Deutsche Nationalbibliothek: The Deutsche Nationalbibliothek lists this publication in the Deutsche Nationalbibliografie; detailed bibliographic data are available in the Internet at http://dnb.d-nb.de.
Any brand names and product names mentioned in this book are subject to trademark, brand or patent protection and are trademarks or registered trademarks of their respective holders. The use of brand names, product names, common names, trade names, product descriptions etc. even without a particular marking in this works is in no way to be construed to mean that such names may be regarded as unrestricted in respect of trademark and brand protection legislation and could thus be used by anyone.

Cover image: www.ingimage.com

Publisher: Südwestdeutscher Verlag für Hochschulschriften GmbH & Co. KG
Heinrich-Böcking-Str. 6-8, 66121 Saarbrücken, Germany
Phone +49 681 37 20 271-1, Fax +49 681 37 20 271-0
Email: info@svh-verlag.de

Printed in the U.S.A.
Printed in the U.K. by (see last page)
ISBN: 978-3-8381-3341-6

Copyright © 2012 by the author and Südwestdeutscher Verlag für Hochschulschriften GmbH & Co. KG and licensors
All rights reserved. Saarbrücken 2012

Table of Contents

Zusammenfassung	II
Summary	IV
1. Introduction	1
1.1. Chitin	1
1.2. Chitin degradation	2
1.3. Chitinases	6
1.4. Detection methods of bacterial chitinases	9
1.5. Distribution of bacterial chitinases	13
1.6. Archaeal chitinases	15
1.7. Goals of the work	18
2. Material and Methods	21
2.1. Origin of strains used in this study	21
2.2. Isolation of chitin degrading bacteria	22
2.3. Identification of archaeal chitinase genes	23
2.4. Primer design	23
2.5. DNA extraction	24
2.6. PCR	24
2.7. Sequencing	27
2.8. Sequence analyses	29
2.9. Cloning of archaeal chitinases	30
2.10. Chitinase activity	33
3. Results	35
3.1. Establishment of a screening panel for chitin degrading microorganisms	36
3.2. Screening of natural occurring bacteria and strain collections for chitinolytic activity	46
3.3. Detection, isolation and characterisation of archaeal chitinases	59
4. Discussion	71
4.1. The chitinase test panel	71
4.2. Screening of new environmental isolates and strains from collections	76
4.3. Archaeal chitinases	83
5. Future perspectives	89
6. Bibliography	91
7. Appendix	105

Zusammenfassung

Chitin ist nach Zellulose das zweithäufigste Biopolymer auf der Erde. Sein Vorkommen ist enorm. Schätzungen gehen von einer Jahresproduktion von bis zu 10^{11} Tonnen aus. Chitin besteht aus N-Acetyl-Glukosaminuntereinheiten, die miteinander β-1,4-glykosidisch verknüpft sind. Der Abbau von Chitin ist vor allem in den Ozeanen ein sehr wichtiger Vorgang, um einen fortwährenden Nachschub an Kohlenstoff und Stickstoff sicher zu stellen. Aufgrund seiner Struktur ist Chitin sehr widerstandsfähig gegenüber dem physikochemischen Abbau. Es wird hauptsächlich von Mikroorganismen biologisch abgebaut. Bisher sind drei Abbauwege bekannt, die sich verschiedener Enzyme bedienen. Ein sehr wichtiges Enzym ist hierbei die Chitinase, die sich sowohl in Bakterien als auch in Pilzen und Archaea findet. Chitinasen hydrolisieren die β-1,4-glykosidische Verknüpfung zwischen den N-Acetyl-Glukosaminuntereinheiten. Sie werden nicht nur zur Nahrungsaufnahme, sondern auch beim Häuten von Arthropoden und bei der Immunabwehr höherer Organismen verwendet. Chitinasen werden in ökologischen Untersuchungen als Anzeiger für chitinolytische Mikroorganismen eingesetzt. Hierbei wird meistens molekularbiologisch nach dem genetischen Fingerabdruck von Chitinasen gesucht oder die jeweiligen Mikroorganismen werden auf Chitin kultiviert, um fest zu stellen, ob Chitin verwertet werden kann. Hierbei wurde bei den meisten Studien in der Vergangenheit nur eine der beiden Herangehensweisen verwendet. Auch der direkte Nachweis des Enzymes fehlt in den meisten Studien. Außerdem wurden bisherige Arbeiten vor allem auf Bakterien und Pilze fokussiert, wobei Archaea kaum berücksichtigt wurden. Um einen umfassenderen Ansatz zu finden, wurde in dieser Arbeit ein dreistufiges Screening Panel etabliert und getestet. Das Test-Panel besteht aus der Isolation und Kultivierung von Mikroorganismen auf Chitin als einziger Kohlenstoff- und Stickstoffquelle, dem molekularen Screening der kultivierten Isolate auf das Vorhandensein eines genetischen Chitinasemotivs, sowie der Evaluation der entsprechenden Chitinaseaktivität. Das etablierte Verfahren wurde genutzt, um verschiedene Lebensräume im Meer in Bezug auf die

Zusammenfassung

Chitin-abbauenden Bakterien zu vergleichen. Hierzu dienten Proben von Sedimenten aus dem Mittelmeer und der Oberfläche von Ostseegarnelen. Außerdem wurde Bakterien (*Actinomyceten* und Isolate von Bryozoen) aus der KiWiZ-Stammsammlung untersucht. Insgesamt wurden 145 Bakterienstämme in dieser Arbeit auf ihre chitinolytischen Eigenschaften hin untersucht. Weiterhin wurde in dieser Arbeit das erste crenarchaeelle Chitinase-Gen in *Sulfolobus tokodaii* identifiziert, in *E. coli* exprimiert und als aktive Chitinase erstmals beschrieben. Das Chitinase-Gen des Euryarchaeons *Halobacterium salinarum* wurde ebenfalls zum ersten Mal in *E. coli* exprimiert und das Genprodukt als aktive Chitinase charakterisiert.

Summary

Chitin is after cellulose the second most abundant biopolymer on earth. It's production is enormous, with estimates of up to 10^{11} tons for both the annual production and the steady-state amount. Chitin consists of β-1,4 glycosidic bonded N-acetyl-glucosamine subunits. It's degradation is especially in the oceans an important step to ensure the continuous availability of carbon and nitrogen. Chitin is very resilient to physicochemical degradation due to its structure. It is mainly biodegraded by microorganisms. Until now three degradation pathways are know, utilising different enzymes. One very important enzyme found in the biodegradation pathways of bacteria, fungi and archaea is the chitinase. Chitinases hydrolyse the β-1,4 glycosidic bond between the N-acetyl-glucosamine subunits. This enzyme is used not only for the recovery of nutrients in microorganisms, it plays also a major role in moulting of arthropods and is utilised in defence mechanisms of higher organisms.

This important enzyme, as proxy for chitinolytic activity, is detected with molecular methods and cultivation based approaches. Most of the studies detecting chitinases do either test for the genetic capability or the growth capability of the respective microorganisms on chitin. Moreover, direct proof of the the chitinolytic enzyme itself is lacking in many studies. Until now a more comprehensive chitinase test panel, combining cultivation and molecular screening of the cultivated strains for their genetic capabilities has not been implemented yet. Furthermore, the search for chitinolytic organisms was mainly focused on bacteria and fungi, but almost no chitin degrading archaea were detected until now.

Within this study a novel three step chitinase test panel was established and tested, consisting of isolation and cultivation of microorganisms on chitin as sole carbon and nitrogen source, the molecular screening of the cultivated strains and the evaluation of the respective chitinase activity. This approach was used to investigate bacteria isolated from different marine microbial communities (Mediterranean Deep Sea sediments and Baltic Sea shrimp carapaces). In addition, bacterial strains (bryozoan derived isolates and actinomycetes) of the KiWiZ strain collection were also investigated. In total, 145 bacterial strains were investigated in this study. Furthermore, the first crenarchaeal chitinase gene from *Sulfolobus tokodaii* was detected, expressed in *E. coli* and the resulting chitinase was described. In addition, the chitinase gene of the halophilic euryarchaeon *Halobacterium salinarum* was expressed for the first time in *E. coli*.

1. Introduction

1.1. Chitin

Chitin is after cellulose the second most abundant biopolymer on earth. It is the main component in the exoskeletons of arthropods like crustaceans, spiders and insects and the cell wall of many fungi. The radulas of molluscs and the beaks and cuttlebones of cephalopods consist of chitin and it is also found in the egg-shells and the oesophagus of nematodes (Jolles & Muzzarelli 1999). Recently it was also discovered as structural element in sponges (Ehrlich et al. 2010). In all these organisms, chitin is used due to its robust but yet flexible nature to strengthen structures or act as skeleton itself. Chitin production is enormous, with estimates of 10^{10} to 10^{11} tons for both the annual production and the steady-state amount (Gooday 1990b, Keyhani & Roseman 1999, Patil et al. 2000) in the marine environment alone.

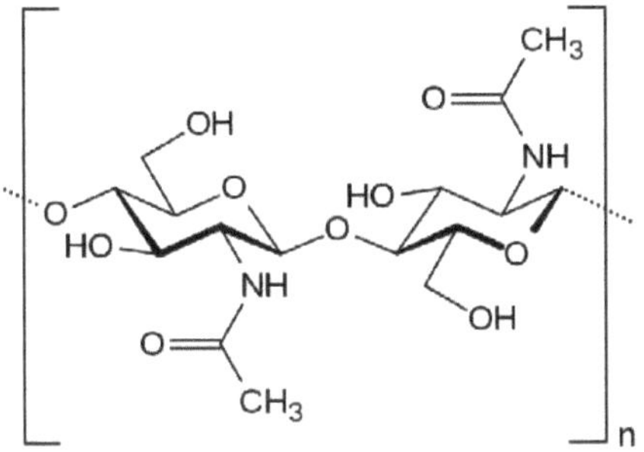

Figure 1.1: Detail of the chemical structure of chitin, depicting two β-1,4 glycosidic bonded N-acetyl-glucosamine subunits.

Chemically, chitin is a polysaccharide, consisting of β-1,4 glycosidic bonded N-acetyl-glucosamine subunits in various grades of acetylation (Fig. 1.1). Its chemical structure is very similar to cellulose, however, the additional acetyl-amine group strengthens the intramolecular hydrogen bonds in the polymer and increases durability and strength of chitin compared to cellulose. Until now three hydrogen-bonded crystalline chitin forms were described: α-chitin with antiparallel chains, β-chitin

Introduction

with parallel chains and γ-chitin with a three chain unit; two "up" and one "down" (Blackwell 1988). The pure chitin polymer does not occur in the natural environment, except in diatom spines. In all other organisms chitin microfibrils are cross-linked and immersed in a matrix of proteins and other polysaccharides, which allows the matrix to resist tensions and gives elasticity. In return, the cementing compounds protect chitin from chemical attacks and keep the microfibrils separate, thereby preventing fracture and providing support to counteract tensions (Ruiz-Herrera & Martinez-Espinoza 1999). The crystalline chitin form embedded in this matrix also influences the physical properties and is crucial for the biological function. This can be observed in the squid *Loligo* sp., having α-chitin in its tough beak, β-chitin in its rigid pen (cuttlebone) and γ-chitin in its flexible stomach lining (Gooday 1990b).

1.2. Chitin degradation

Chitin is practically insoluble in water, diluted acids, diluted and concentrated alkalis, alcohols and other organic solvents. It is only soluble in concentrated hydrochloric acid, sulphuric acid, 78-97 % phosphoric acid and anhydrous formic acid (O'Neil 2006). In addition, the matrix in which chitin is embedded in its natural form, does prevent chemical attacks, further impeding its physicochemical degradation. However, due to the high carbon and nitrogen content of chitin, its degradation is an extremely important step in nutrient cycling, especially in the oceans (Kirchman & White 1999). Without the degradation of chitin, carbon and nitrogen would be depleted very fast. Hence, the biological degradation of chitin is very important (Poulicek et al. 1998).

Chitin biodegradation pathways

Until now only three biodegradation pathways of chitin have been described (Fig. 1.2). The three described chitinolytic pathways are:

(A) Chitin is degraded into chitooligosaccharides ($GlcNAc_n$) and into N-acetyl-glucosamine dimers ($GlcNAc_2$) by chitinases (reaction 1). The glucosamine dimers

Introduction

are then hydrolysed by β-N-acetylglucosaminidase (reaction 2) to form N-acetyl-glucosamine (GlcNAc) or to release GlcNAc directly from chitooligosaccharides (Gooday 1990b). Some organisms degrade $GlcNAc_2$ to GlcNAc and GlcNAc-1-phosphate by $GlcNAc_2$ phosphorylase (reaction 3) (Park et al. 2000) or convert the dimer to GlcNAc-6-phosphate-GlcNAc by a $GlcNAc_2$ phosphotransferase system (reaction 4) followed by degradation to GlcNAc and GlcNAc-6-phosphate by 6-phospho-β-glucosaminidase (reaction 5) (Keyhani et al. 2000). This pathway was found in bacteria and fungi.

(B) Another pathway for chitin degradation by bacteria and fungi is proposed to occur through deacetylation of chitin by chitin deacetylase (reaction 6). The resulting deacetylated chitin, chitosan, is then degraded to glucosamine (GlcN) by chitosanase (reaction 7) in cooperation with exo-β-D-glucosaminidase (reaction 8) (Gooday 1990b).

(C) The third pathway has until now only been discovered in the archaeon *Thermococcus kodakaraensis*. In this pathway chitin is degraded by a chitinase to $GlcNAc_2$ (reaction 1). This dimer is partially deacetylated by $GlcNAc_2$ deacetylase to the disaccharide GlcN-GlcNAc (reaction 9). This product is hydrolysed to GlcN and GlcNAc monomers by exo-β-D- glucosaminidase (reaction 8) (Tanaka et al. 2003). Finally, GlcNAc is deacetylated by $GlcNAc_2$ deacetylase (reaction 9), resulting in the complete conversion of chitin to GlcN monomers (Tanaka et al. 2004).

Just recently, an oxidative enzyme has been discovered, enhancing the conversion chitin. This chitin-binding-protein (CBP) acts as an enzyme on the surface of crystalline chitin, inducing chain breaks in the chitin and thereby alleviating access for other enzymes (Eijsink et al. 2010). In all cases, the resulting end products can then be phosphorylated and the acetyl and amino groups can be sequentially removed to generate fructose-6-phosphate, which can enter the glycolytic pathway (Boulanger et al. 2010).

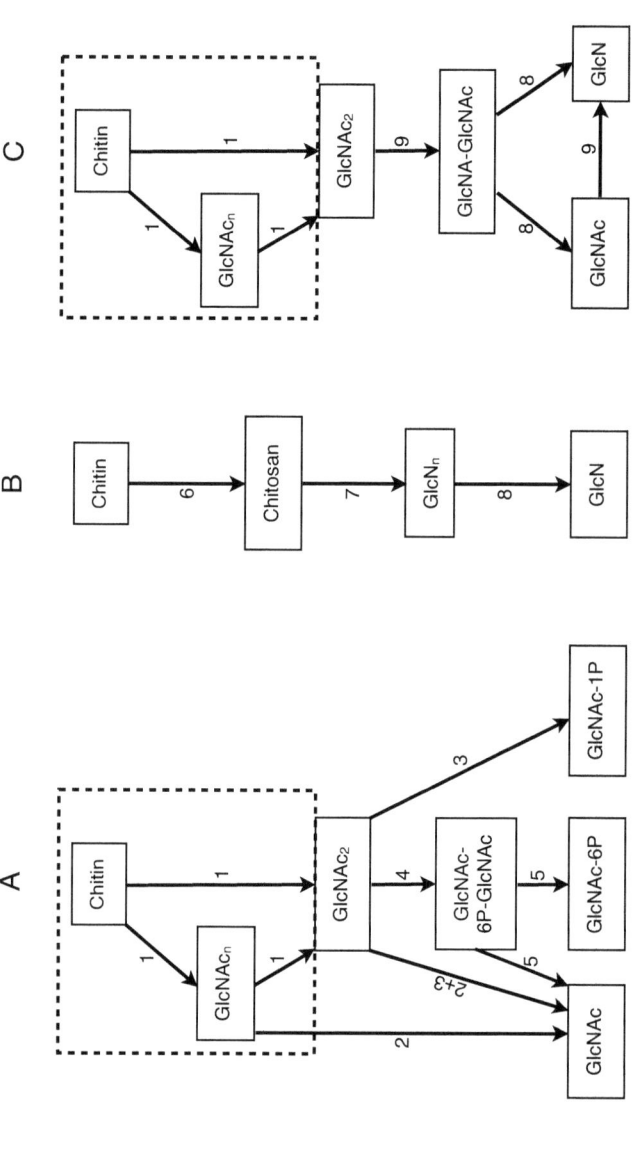

Figure 1.2: Chitin catabolic pathways from chitin to monosaccharides. A and B: chitinolytic pathways known from bacteria and fungi. C: chitinolytic pathway found in the archaeon *Thermococcus kodakaraensis* KOD1. Numbers represent different enzymes: (1) Chitinase, (2) β-N acetylglucosaminidase, (3) GlcNAc$_2$ phosphorylase, (4) GlcNAc$_2$ phosphotransferase system, (5) 6-phospho-β-glucosaminidase, (6) chitin deacetylase, (7) chitosanase, (8) exo-β-D-glucosaminidase, (9) GlcNAc$_2$ deacetylase. Abbreviations of intermediates: (GlcNAc$_n$) N-acetylchitooligosaccharide, (GlcN$_n$) chitooligosaccharide, (GlcNAc$_2$) diacetylchitobiose, (GlcN$_2$) chitobiose, (GlcNAc1P) GlcNAc-1-phosphate, (GlcNAc6P) GlcNAc-6-phosphate. Dashed boxes indicate action of chitinases and highlight the enzymatic pathways investigated in this study.

Introduction

Functions of chitinolysis

The chitin biodegradation pathways are mainly used to obtain nutrients. The majority of chitin degraders for nutritional purposes are found among the bacteria and fungi (Gooday 1990b). Bacteria are reported to dominate the chitinolytic community in the oceans, ocean sediments and freshwater environments. Characteristic genera are *Vibrio, Photobacterium, Serratia, Chromobacterium, Pseudomonas, Flavobacterium, Bacillus, Cytophapa* and *Actinomycetes*, with the latter two foremost found in sediments (Gooday 1990a). Fungi on the other hand are reported to dominate the chitinolytic community in terrestrial habitats. Fungal genera with most common chitinolytic specimen are *Aspergillus, Trichoderma, Verticillium, Thielavia, Penicillium* and *Humicola* (Gooday 1990a). Also symbiotic interactions have been proposed, e.g. in the guts of whales to help digest the carapaces of krill (Gooday 1990b). However this is still critically discussed, as chitinolytic enzymes have also been detected in fish, plants and other higher organism. Hence, unlike the degradation of the chemically very similar cellulose, chitin degradation in higher organisms for nutritional purposes does not always require a symbiotic chitinolytic microbial community (Gooday 1990a). Bacteria and fungi can also be found as pathogens of chitin producing organisms, using chitinolytic enzymes either to access other substrates, or leaching the chitin of the host directly. In addition to chitin degradation for nutritional purposes, chitinolysis plays also an important role in all organisms that synthesise chitin. These organisms need chitinolysis for autolytic and morphological reasons. This can be observed for example in fungi during germination and spore formation: The autolysis of the fungal cell wall at the end of spore maturation ensures the release of the spores (Jolles & Muzzarelli 1999). Examples for the morphological necessity of chitinolysis are moulting in arthropods or apical growth and branching in fungal hyphae (Karlsson & Stenlid 2008). Chitinolytic enzymes, especially chitinases, are also involved in the immune response of plants (Salzer et al. 2000), where they are classified as pathogenesis-related proteins and are utilised for example to degrade infecting fungal hyphae (Kasprzewska 2003). A similar function has also been reported for mammals

(Funkhouser & Aronson 2007), where functional chitinases have been found to be expressed and excreted in the guts, during pathogen attacks and also during allergic reactions (Bussink et al. 2006). Just recently chitinases were reported to play a potential role in the pathogenesis of asthma (Shuhui et al. 2009).

When comparing the different chitinolytic pathways, especially chitinases are found as key enzymes in bacterial, fungal and archaea chitinolysis. Chitinases initiate chitin degradation and were found also in pathogenic bacteria and fungi, in plants and even mammals, where they play important roles apart from nutritional purposes. Chitinases were put in the focus of this work, as crucial enzymes of bacteria, archaea, fungi and higher organisms.

1.3. Chitinases

Chitinases are classified according to their function of hydrolysing chitin in the enzyme classification system (EC). Within this system, enzymes are attributed to different enzyme classes and subclasses according to their capability to convert substrates. Chitinases belong to the enzyme class EC 3.2.4.14. As the physical form of an enzyme does influence the substrate specificity, enzymes can also be classified according to their amino acid sequence. As chitinases hydrolyse glycosidic bonds, they are attributed to the amino acid sequence homology based system of the glycoside hydrolase superfamily (Henrissat & Davies 1997). Within this system, chitinases belong to the glycoside hydrolase (GH) families 18 and 19 (Henrissat 1991).

Introduction

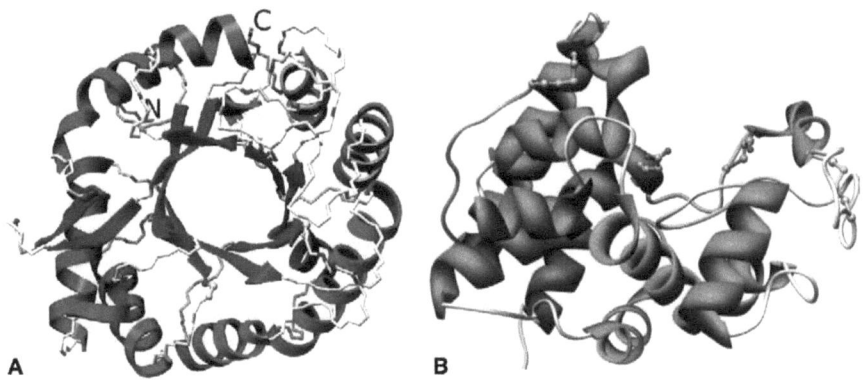

Figure 1.3: Structure of the active centre of glycoside hydrolase family 18 and 19. (A) GH family 18. Part of the *Pyrococcus furiosus* chitinase showing the active site with the eight α-helices, depicted in red and the eight β-strands, depicted in blue. C denotes C-terminal side, N denotes N-terminal side (Nakamura et al. 2007). (B) GH family 19 active centre of the *Streptomyces griseus* chitinase C. α-helices are shown in green (Kezuka et al. 2006).

Glycoside hydrolase family 18 (GH18)

GH family 18 harbours mainly chitinases from bacteria (LeCleir et al. 2004), but also includes chitinases from eukaryotes, viruses and archaea (Karlsson & Stenlid 2009). This family was further divided into the three subfamilies A, B and C according to differences in the amino acid sequences (Suzuki et al. 1999). The GH18 subfamily A was proposed to be prevalent among the bacterial chitinases (Metcalfe et al. 2002). GH family 18 chitinases consist in general of one catalytic domain and one or more chitin binding domains (Kezuka et al. 2006). The enzyme family is characterised by its barrel structure (TIM Barrel) consisting of eight α-helices and eight parallel β-strands that alternate along the peptide backbone (Wierenga 2001) (Fig. 1.3). It operates through a retaining mechanism, in which a β-linked polymer is cleaved to release the β-anomer. The catalytic residue has been identified experimentally to be a glutamic acid (Tsuji et al. 2010).

Glycoside hydrolase family 19 (GH19)

GH family 19 comprises mainly plant chitinases but also enzymes from bacteria, other eukaryotes and viruses (Kezuka et al. 2006). The members of this enzyme family have a bilobal structure with a high content of α-helices (Fig. 1.3). This bilobal structure is very variable and until now 20 different core motives were identified based on the amino acid sequence (Prakash et al. 2010). Family 19 enzymes generally operate through an inverting mechanism, in which a β-linked polymer is cleaved to release the α-anomer. The catalytic residue of the enzyme class has not been verified yet.

As already mentioned, chitinases can be found in all domains of life. Most of the chitinases belong to the bacteria while almost no chitinases were detected until now in the domain of archaea. The focus of this research was further narrowed to bacterial and archaeal chitinases.

Introduction

1.4. Detection methods of bacterial chitinases

The search for bacterial chitinases is a search for chitin degraders in the environment. This search is based on direct and indirect proof of chitinase activity. The direct proof shows the chitinase activity of an enzyme, while the indirect proof shows the effects of chitin degradation and it can be assumed that these effects are based on the presence of chitinolytic enzymes. In general, two main approaches are used for the detection of chitin degraders, regardless of direct or indirect proof of chitinolysis. These are the culture dependent and the culture independent methods.

Culture dependent methods

Classical isolation of microorganisms on specific (chitin containing) media is the basis of culture dependent methods. Various modifications are used. A commonly used method is the *in situ* enrichment, incubating chitin as substrate in the environment (Metcalfe et al. 2002, Hobel et al. 2005, Kublanov et al. 2008), followed by the cultivation of the resulting microbial community under laboratory conditions and the isolation of chitinolytic strains.

To assess the chitinolytic capabilities of an isolated strain four main techniques are utilised:

- Observation of growth behaviour on chitin: The bacterial strains are grown on a minimal medium containing chitin carbon and nitrogen source. This medium is turbid and the chitin is clearly visible. During the growth of the different strains the chitin contained in the agar medium is degraded by the bacterial chitinases, thereby forming clearing zones around the strains. By the occurrence of these clearing zones the chitin degrading strains can be distinguished from other bacterial strains. This is an indirect proof for the presence of chitinases, as the enzymes responsible for the degradation of the chitin within the medium are not isolated and tested directly.

Introduction

- Purification and measurement of chitinase activity of the respective enzyme: The proteins of the respective bacterial strain are extracted, separated, purified and tested for their chitinolytic potential (as described below). This is a direct proof of chitinase activity, as the enzyme itself is used as sample.

- Measurement of products from digestion of chitin oligomers: Chitin oligomers are labelled with a fluorogenic dye that is glycosidically bonded. The labelled analogues of chitin oligomers are incubated with the respective microorganisms or protein extracts of the microorganisms and the release of the chromophore after enzymatic hydrolysis is measured. Depending on the used sample, protein extract or microorganisms, this approach is either direct (used with protein) or indirect (used with microorganisms).

- Measurement of products from digestion of chitin: This method is very similar to the latter one. The difference is the used substrate and the detection of chitinolysis. During the hydrolysis of the glycosidic bond between the chitin subunits a reducing sugar end is created. This increase in reducing ends is measured. The benefit of this method is that the used chitin has not to be chemically modified and thereby mirrors the natural occurring conditions. As mentioned above, this approach is either direct, when used with protein as sample, or indirect, when used with microorganisms as sample.

These four methods are used to characterise chitinolytic bacterial communities. However, they have their limitations. For example, it is known that not all microorganisms can be grown under laboratory conditions, or the cultivation techniques can yield different results, as the mechanisms involved in the detection of chitinolysis are different: When using the clearing zone method, chitinases have to be excreted and diffuse into the medium to reach the chitin. Hence, chitinases anchored to the membrane are rarely detected. In addition, this method is very time consuming, since growth on chitin minimal media is slow and therefore the formation of clearing zones is also very slow. Furthermore, chitin degradation can also be accomplished in rare cases by other enzymes (see Fig. 1.2). However, the use of a solid chitin minimal medium allows the isolation and separation of bacterial strains while simultaneously detecting chitinases. When using fluorogenic labelled analogues of chitin as substrate the chain length of the used chitin analogue is crucial. Natural occurring chitin has a high molecular weight and is not as easily accessible for enzymes as short oligomeric chains. Hence, the capability to cleave glycosidic bonds is detected by this method and not necessarily the capability to degrade high molecular weight chitin. The same holds true for the detection of reducing ends: When using high molecular weight chitin, chitinases are detected. When using oligomers, the capability of the respective enzyme to cleave glycosidic bonds is detected. The differences in the detection mechanism are directly reflected in the results. This can be seen for example in the estimates of chitin degrading bacteria within different communities. Cottrell (1999) reported that the percentage of bacteria supposed to be chitin degraders is estimated to be 10 % within a bacterial community when using the clearing zone approach; the fluorogenic labelled analogues approach lead to a proportion of 90 % of chitin degraders within another comparable bacterial community, which is very unlikely. These differences and limitations have to be considered, when choosing a chitinase detection method. The method of choice for many scientists is the measurement of the degradation of short chain, fluorogenic labelled analogues, as it is very fast, reliable and easy to perform (Hood 1991). During this work it was decided to use chitin minimal medium for the initial isolation, consecutive cultivation and initial chitinase detection.

Culture independent methods

It is known that not all microorganisms can be readily and easily cultivated under laboratory conditions. To further elucidate the *in situ* bacterial chitin degrading communities, molecular techniques are used. Genetic material of the respective sample is isolated and subjected to the amplification of genes encoding for e.g. chitinases via PCR with specific primers. Two general primer designs are used:

Introduction

- Genus specific primers that are designed according to known chitinase sequence (LeCleir et al. 2004). Available chitinase sequences of the different bacterial strains from the same genus are aligned to generate a consensus sequence. With this consensus sequence chitinase primers are generated for the genetic detection of chitinases. Depending on the chosen DNA template the resulting primers can be genus specific or detect a broader range of chitinases.

- Motif specific primers that are designed according to known chitinase motifs of glycoside hydrolase families (Hobel et al. 2005). In contrast to the latter primer design, chitinase sequences from various bacterial genera belonging to the same glycoside hydrolase family (e.g. GH18) are aligned and a consensus sequence is generated to design motif specific degenerated primers for the detection of chitinases.

Both primers work culture independent and rely on the availability of chitinase sequence data. As motif specific primers do cover a broader range of the bacterial diversity as compared to genus specific primers and two degenerated primer sets for the glycoside hydrolase family 18 A motif have been already designed, tested and used to describe natural occurring chitinolytic communities (Hobel et al. 2005), these primers were used in this study to further complement the cultivation based approach.

In addition to screening bacterial communities with primers, searches can also be conducted purely with bioinformatic techniques. Gene sequences of microorganisms are screened for the presence of chitinase specific motifs. This molecular mining is conducted completely *in silico* and gives information about potential chitinases purely on the basis of sequence information. Chitinases detected in this way can be heterologously expressed, purified and tested for their actual chitinase activity. This method should be treated with caution, as the presence of chitinase motifs themselves shows a chitinolytic potential of the respective organism, but only the expressed and active protein proves directly the chitinolytic activity. This molecular approach was also used to complement the cultivation based approach and utilised for the search for archaeal chitinases and linked with the overexpression and characterisation of the potential chitinase.

1.5. Distribution of bacterial chitinases

Bacterial chitinase sequences have been detected in a range of diverse environments. Their distribution was shown and studied in various marine and non-marine environments, including alkaline soils, where sequences of alkalophilic *Streptomycetes* were found (Tsujibo et al. 2003); Antarctic lake sediments, with sequences that were attributed to the *Gammaproteobacteria, Actinobacteria* and the CFB-group, but interestingly no sequences of psychrophilic bacteria (Xiao et al. 2005); estuarine water and sediments, which were dominated by sequences of *Vibrio, Serratia, Bacillus* and *Aeromonas* species (Ramaiah et al. 2000); freshwater and saline lakes and the central Arctic Ocean, dominated by *Gammaproteobacteria* and Gram-positive bacterial sequences (LeCleir et al. 2004) and deep sea sediments, with a majority of *Serratia* sequences (Lian et al. 2007). Also first approaches of combined molecular and cultivation studies have been undertaken. Hobel and coworkers (2005) investigated intertidal hot springs and found sequences of *Firmicutes, Betaproteobacteria, Gammaproteobacteria* and *Actinobacteria* dominating the bacterial chitinolytic community when investigated with genetic methods, while the cultured chitinolytic bacterial community obtained on chitin minimal medium was dominated by *Proteobacteria, Firmicutes* and *Actinomycetes*. Another kind of combined cultivation and genetic screening approach was used to investigate upland pastures. Bags filled with chitin were buried and incubated in the soil and the resulting microbial community was screened using 16S-rDNA denaturing gradient gel electrophoresis (DGGE). The DGGE patterns of the chitin bag enriched bacterial soil community showed more bands as the DGGE pattern of not enriched bacterial soil communities, indicating an enrichment of bacteria in chitin treated soils. Furthermore, the genetic investigation showed a dominance of actinobacterial chitinase sequences in the chitin enriched soils. This result was mirrored in the count of colony forming actinobacterial units, conducted with chitin amended actinomycetes medium. Pure cultivation approaches have also been used to obtain chitinolytic bacteria. As reviewed by Gooday (1990a),

Introduction

chitin degrading bacteria have been isolated on chitin amended media from the oceans, ocean sediments, freshwater environments and terrestrial sediments. The isolated strains were mainly attributed to the genera *Vibrio, Photobacterium, Serratia, Chromobacterium, Pseudomonas, Flavobacterium, Bacillus, Cytophapa* and *Actinomycetes*.

These works showed the broad distribution of chitinolytic bacteria in various habitats and also in the different phylogenetic genera. It is noticeable that molecular and cultivation studies predominate, but studies combining both approaches are rare. When both approaches are used, neither is tested if a bacterial strain with the genetic potential to degrade chitin does degrade chitin, nor is tested whether the strains growing on solid chitin medium use chitin as only carbon and nitrogen or if the respective bacterium does posses a known chitinase sequence. It is in general not tested until now whether the detected chitinases are excreted into the surrounding medium or not. Within this work emphasis was also given to the task of developing a more comprehensive approach of chitinase detection in microorganisms aiming to describe chitinolytic bacterial communities.

Introduction

1.6. Archaeal chitinases

The second microorganism group in the focus of this work is the domain of the archaea, as only little is known about chitinases from this domain. Until now only ten euryarchaeal chitinases were detected so far (Tab 1.1). In contrast to bacterial chitinases, archaeal chitinases are mainly found by searching in the genomes of sequenced archaeal strains for key motifs of known chitinases. Thus, their actual chitin degrading capabilities have mostly not been elucidated. Until now, enzymatic characterisation including proof of chitinase activity has been accomplished for four organisms:

- In *Thermococcus chitinophagus,* the only archaeon shown to grow directly on chitin as carbon and energy source so far (Huber et al. 1995), a chitinase bound to the outer side of the cell membrane was detected. Its optimal activity was reported to be at 70 °C and pH 7. The chitinase was highly thermostable and showed no inhibition by allosamidin. In addition, it was resistant to denaturation by urea and SDS (Andronopoulou & Vorgias 2004).

- Tanaka et al. (Tanaka et al. 1999) cloned and overexpressed the chitinase gene of *Thermococcus kodakaraensis* KOD1. The enzyme showed optimal activity at 85 °C and pH 5. It had dual active sites (GH18 A and GH18 C) and three substrate binding sites, according to the amino acid sequence and genetic deletion experiments (Tanaka et al. 2001).

- The genome of *Pyrococcus furiosus* supposedly comprised two chitinases, belonging to the subfamilies GH18A and B. Both were cloned and expressed in *E. coli*. They showed pH optima at pH 6 with thermal optima between 90 °C and 95 °C. Furthermore, these two chitinases acted synergistically when incubated together on colloidal chitin, resulting in a fivefold increase compared to incubation with only one of the two chitinases (Gao et al. 2003). However, the subsequent DNA sequence analysis showed that the two genes were formed as a result of a nucleotide insertion, causing a frame shift (Nakamura et al. 2007). After removal of the in-

Introduction

serted nucleotide, an artificial recombinant chitinase was expressed by Oku et al. (2006), resulting in a 40 fold increase in chitinase activity.

- Hatori et al. (2006) found a putative chitinase gene in the genome of *Halobacterium* sp. strain NRC-1 and expressed it in the extremely halophilic archaeon *Haloarcula japonica* strain TR-1. The enzyme was reported to be halophilic, with an optimal activity at about 1 M NaCl. The activity was retained at salt concentrations ranging up to approximately 5 M NaCl. In addition, the enzyme was insensitive to DMSO concentrations of up to 30 % (v/v). Unfortunately, no kinetic data were given by the authors.

Additionally, six sequences of putative chitinases were annotated, all from euryarchaeal organisms. Until now, no chitinases or the respective genes have been described or annotated within the crenarchaeal group.

The knowledge concerning archaeal chitinases is mainly based on molecular works. *In situ* research, isolation of archaeal chitin degraders and proof of the chitin degrading capabilities is still very scarce (Tab 1.1). Hence, to broaden the existing knowledge concerning archaeal chitinases is another focus of this work.

Introduction

Table 1.1: List of archaeal chitinases described until 2012.

Organism	Enzyme family	Type of characterization	Reference
Thermococcus chitinophagus	GH18	Native purification, activity confirmed, characterized	Genbank Acc. No. AAR13021.1 (Andronopoulou & Vorgias 2004)
Thermococcus kodakaraensis KOD1	GH18	Recombinant enzyme, activity confirmed, characterized	Genbank Acc. No. BAD85954.1 (Tanaka et al. 1999)
Pyrococcus furiosus	GH18	Recombinant enzyme, activity confirmed, characterized	Genbank Acc. No. AAL81357.1 (Oku & Ishikawa 2006)
Halobacterium sp. strain NRC-1	GH18	Recombinant enzyme, activity confirmed	Genbank Acc. No. AAG19274.1 (Hatori et al. 2006)
Halobacterium salinarum	GH18	Genome annotation	Genbank Acc. No. CAP13543.1 (Pfeiffer et al. 2008)
Halomicrobium mukohataei DSM 12286	GH18	Genome annotation	Genbank Acc. No. ACV49028.1 (Tindall et al. 2009)
Haloterrigena turkmenica DSM 5511	GH18	Genome annotation	Genbank Acc. No. ADB61056.1 (Saunders et al. 2010)
Methanoplanus petrolearius DSM 11571	GH18	Genome annotation	Genbank Acc. No. ADN37298.1
Candidatus Korarchaeum cryptofilum OPF8	GH18	Genome annotation	Genbank Acc. No. ACB07477.1 (Elkins et al. 2008)
Candidatus Methanoregula boonei 6A8	GH18	Genome annotation	Genbank Acc. No. ABS56694.1 (Bräuer et al. 2010)

1.7. Goals of the work

This work focuses on bacterial and archaeal chitin degraders and chitinases with the following goals:

Establishment of a more comprehensive chitinase test panel

A novel approach of combined cultivation and molecular screening for chitinase detection in microorganisms, also focusing on the presence of excreted chitinases shall be established and validated in this work, to further increase the knowledge concerning the presence of chitinase genes, their expression and excretion and the use of chitin as sole carbon and nitrogen source.

Screening of natural occurring bacterial communities and strain collections for chitinolytic activity

The established chitinase test panel shall be used to isolate marine bacterial chitin degraders, elucidate the chitinolytic activity and potential of natural occurring bacterial communities and to screen strains collections for chitinolytic bacteria.

Detection, overexpression and characterisation of archaeal chitinases

Within this work the still scarce knowledge concerning the existence of archaeal chitin degraders and chitinases shall be broadened, using molecular techniques, to allow for further works concerning archaeal chitin degraders and elucidation of the evolution of chitinases.

To elucidate these goals molecular and cultivation based techniques are used. The goals of this work are interconnected as shown in figure 1.4 (Fig. 1.4).

Introduction

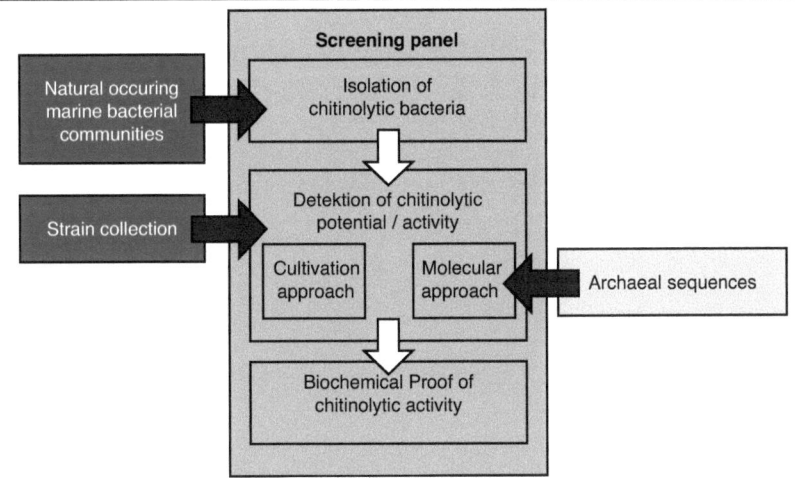

Figure 1.4: Schematic overview of the goals of this work. Grey: screening panel. Dark grey: screening of natural occurring bacterial communities and strain collections. Light grey: detection of archaeal chitinases. White arrows depict working flow within the screening panel. Black arrows depict different input points to start the screening process during this study.

2. Material and Methods

2.1. Origin of strains used in this study

Within this work three different sources of bacterial and archaeal strains were used. The strains isolated during sampling campaigns were used for the establishment and validation of the screening panel and the elucidation of the distribution of chitin degrading bacteria; the strains from the KiWiZ strain collection were used to screen for chitinolytic bacteria and the obtained cultures from the DSMZ were used for the elucidation of archaeal chitinases.

Sampling campaigns

Bacterial strains were isolated from crustaceans and sediment cores that were obtained by SCUBA diving and corers depths of up to 4400 m. The samples originated from different habitats in the Baltic and Mediterranean Sea (Tab. 2.1). Samples were either transferred within 2 h to the laboratory and kept at 4 °C until processing or directly processed at the sampling site.

Table 2.1: Sample origin.

Sampling site	Sample	Method / Depth
Kiel fjord, Baltic Sea	*Palaemon adspersus, Carcinus maenas*	SCUBA / 5 m - 8 m
Ierapetra Basin, eastern Mediterranean	sediment core	Corer / 4400 m
Herodotus plain, eastern Mediterranean	sediment core	Corer / 2800 m
Palinuro Seamound, western Mediterranean	sediment core	Corer / 650 m

Material and Methods

The KiWiZ strain collection

In addition to the obtained strains from the sampling campaigns, marine *Actinomycetes* strains (isolated by Jutta Wiese) and strains isolated from various bryozoans (isolated by Herwig Heindel) were obtained from the KiWiZ strain collection.

The German Collection of Microorganisms and Cell Cultures

The archaeal strains *Sulfolobus tokodaii* strain DSM 16993 (Suzuki et al. 2002) and *Halobacterium salinarum* strain DSM 3754 (Harrison & Kennedy 1922) were obtained from the German Collection of Microorganisms and Cell Cultures (DSMZ) as vacuum dried cultures.

2.2. Isolation of chitin degrading bacteria

Chitin medium

For the isolation and cultivation of chitin degrading bacteria, a solid and a liquid medium were composed. The solid medium, further on referred to as "chitin medium", contained 50 g/l chitin from crab shells (Sigma), 12 g/l Bacto agar, 0.0025 g/l yeast extract (Bacto) and the respective amount of salt (30 g/l NaCl (Roth) for Mediterranean samples, 16 g/l NaCl (Roth) for Baltic Sea samples), or when feasible sterile filtered surrounding water. The liquid medium, further on referred to as "liquid chitin medium" contained 50 g/l chitin (crab shells, Sigma), 0.0025 g/l yeast extract (Bacto) and the respective salt concentration or sterile filtered surrounding water, as already mentioned.

Isolation procedures

Samples were homogenised and diluted in sterile filtered surrounding water in discrete steps to a 10^{-5} dilution. 200 µl of each dilution step were plated onto different growth media (Heindl et al. 2010, Gärtner et al. 2011) and chitin agar plates. Plates were incubated at 28 °C for up to 6 weeks. Growing bacterial strains were singled out by repetitive plating. Pure cultures were transferred to a solid complex medium for ease of handling and cryo conserved at -80 °C using the Cryobank System (Mast Diagnostics) according to the manufacturer.

To exclude the growth of cultures on agar as carbon source, strains were grown in 20 ml of liquid chitin medium at 28 °C and 130 RPM on a shaker for up to seven days. Growth was checked visually by microscopy, as the turbid chitin medium did not allow the use of optical density as measure for cell growth.

2.3. Identification of archaeal chitinase genes

The amino acid sequence of the exochitinase from *Streptomyces olivaceoviridis* (Blaak et al. 1993) was compared with available archaeal sequences in Genbank, the National Centre for Biotechnology Information (NCBI) online database, using the basic local alignment search tool (Altschul et al. 1990) for amino acid sequences BlastP, from NCBI (www.ncbi.nlm.nih.gov).

2.4. Primer design

Primers for the ORF BAB 65950 from *Sulfolobus tokodaii* and ORF CAP 13543 from *Halobacterium salinarum* were designed according to the nucleotide sequence of *Sulfolobus tokodaii* str. 7 (Kawarabayasi 2001) and *Halobacterium* sp. NRC-1 (Ng et al. 2000), respectively. The nucleotide sequences were obtained from the NCBI homepage (www.ncbi.nlm.nih.gov). The primer pair for *Sulfolobus tokodaii* (SkChif and SKChir, Tab. 2.2) was designed not to include the start and stop codon of the respective enzyme, but rather binding upstream of the start codon (SkChif) or downstream of the stop codon (SKChir) respectively. The primer pair for *Halobacterium salinarum* (HBChiNf and HBChiNr, Tab. 2.2) was designed to include the restriction sites for the restriction enzymes Nde I (in the forward primer HBChiN) and BamH1 (in the reverse primer HBChiNr). These primers include the start and stop codon of the putative chitinase gene. After digestion with the respective restriction enzymes the resulting DNA fragment begins with the start codon and ends with the stop codon.

2.5. DNA extraction

Archaea

The vacuum dried cultures were dissolved in 50 µl DNA free water (AppliChem). Total DNA was extracted with the innuPREP Bacteria DNA Kit (Analytik Jena). Crude DNA extract was purified by using the QIAamp DNA Mini Kit (QIAGEN), as described by the manufacturer.

Bacteria

At least 5 mg of bacterial cells were transferred from complex or chitin growth medium into 50 µl DNA free water (AppliChem) respectively. DNA from the isolated strains was obtained by repetitive freezing the suspension at -20 °C, thawing at 100 °C and vortexing for 20 s of the respective bacterial cells for 5 times repetitively. The suspension was finally centrifuged for 5 min at 3500 x g and the supernatant was immediately used as template for the PCR.

2.6. PCR

For amplification of the archaeal chitinase sequences (ORF BAB 65950 from *Sulfolobus tokodaii* and ORF CAP 13543 from *Halobacterium salinarum*), Phusion High-Fidelity DNA polymerase (Finnzyme) was used with concentration of 1 U per reaction and Phusion HF buffer (Finnzyme) in a total volume of 25 µl.

For amplification of bacterial sequences and vector inserts Taq DNA Polymerase (New England BioLabs, final volume of 25µl, 0.5 U per reaction) with the ThermoPol Buffer Kit (New England BioLabs) was used.

All Primers were used at a final concentration of 10 µM. Thermal protocols were conducted in a T1 thermocycler (Whatman Biometra).

Amplification of the archaeal *chi* gene

Amplification of archaeal *chi* genes were conducted using the primer pair SkChif and SkChir for the *Sulfolobus tokodaii* sequence and HBChiNf and HBChiNr for the *Halobacterium salinarum* sequence (Tab. 2.2). Cycler conditions were as follows:

1 cycle	Initial Denaturation	98 °C	30 s
25 cycles	Denaturation	98 °C	10 s
	Annealing	50 °C	30 s
	Elongation	72 °C	90 s
1 cycle	Final Elongation	72 °C	600 s

Colony screening

Clones containing the respective archaeal sequences were screened for positive insert with the primers pQEf and pQEr for the Sulfolobus tokodaii sequence and T7 and T7 term for the *Halobacterium salinarum* sequence, respectively (Tab. 2.2).

Cycler conditions were as follows:

1 cycle	Initial Denaturation	94 °C	120 s
30 cycles	Denaturation	94 °C	40 s
	Annealing	50 °C	40 s
	Elongation	72 °C	60 s
1 cycle	Final Elongation	72 °C	300 s

Amplification of ChiA motif

The core motif of the GH family 18 A was amplified using the primer pairs ChiAf1 with ChiAr1 and ChiAf2 with ChiAr2 (Hobel et al. 2005) (Tab. 2.2).

Cycler conditions were as follows:

1 cycle	Initial Denaturation	95 °C	180 s
35 cycles	Denaturation	95 °C	45 s
	Annealing	55 °C	45 s
	Elongation	72 °C	90 s
1 cycle	Final Elongation	72 °C	300 s

Amplification of 16S rDNA fragment

The almost complete 16S rDNA sequence was obtained with the primers 27f and 1492r (Lane 1991) (Tab. 2.2).

Cycler conditions were as follows:

1 cycle	Initial Denaturation	94 °C	120 s
30 cycles	Denaturation	94 °C	40 s
	Annealing	50 °C	40 s
	Elongation	72 °C	60 s
1 cycle	Final Elongation	72 °C	300 s

DNA size, concentration and purity

The length of the respective DNA fragments was checked on a 2% agarose gel using a 1 kb ladder (Fermentas) as standard. Gels were stained with SYBR safe DNA gel stain (Invitrogen) and evaluated on a UV transilluminator at 366 nm. DNA concentration and purity were determined photometrically with a NanoVue spectrophotometer (GE Healthcare), according to Sambrook et al. (2001).

2.7. Sequencing

DNA sequencing was conducted by the Institute of Clinical Molecular Biology (IKMB) in Kiel according to Sanger et al. (1977). Exonuclease I (Exo I, GE Healthcare, Munich, Germany) and Shrimp Alkaline Phosphatase (SAP, Roche, Grenzach-Wyhlen, Germany) were used for purification of the PCR product. For each reaction 1.5 U of Exo I and 0.3 U of SAP were added to the PCR products and incubated for 15 min at 37°C, followed by heat inactivation of the enzymes for 15 min at 72°C. Sequencing was performed with the BigDye Terminator v1.1 Sequencing Kit (Applied Biosystems, Darmstadt, Germany) in a 3730-DNA-Analyzer (Applied Biosystems, Darmstadt, Germany) as specified by the manufacturer. Primers used for sequencing were: 534r, 342f (Muyzer et al. 1993) and 790f (Thiel et al. 2007) (Tab. 2.1).

Material and Methods

Table 2.2: Primer sequences and origin. Restriction sites are underlined.

Name	Sequence	Origin	Target
SkChif	5'- ATG AAA CGG AAT ACC CTT TTG -3'	This work	Amplification of the putative *Sulfolobus tokodaii* chitinase
SkChir	5'- TTA CCA ATA GTT ATC ACT TCT TTC TC -3'	This work	
HBChiNf	5'- CGC CTC A<u>CA TAT G</u>CC CCA CG -3'	This work, containing restriction site for <u>NdeI</u>	Amplification of the putative *Halobacterium salinarum* chitinase
HBChiNr	5'- GGC GAT <u>GGA TCC</u> TAT CGC TAC -3'	This work, containing the restriction site for <u>Bam H1</u>	
ChiAf1	5'- ACG GCG TGG ACA TCG AYT GGG ART -3'	(Hobel et al. 2005)	Amplification of the glycoside hydrolase family 18 core motive
ChiAr1	5'- CCC AGG CGC CGT AGA RRT CRT AYS -3'	(Hobel et al. 2005)	
ChiAf2	5'- CGT GGA CAT CGA CTG GGA RTW YCC -3'	(Hobel et al. 2005)	
ChiAr2	5'- CCC AGG CGC CGT AGA RRT CRT ARS WCA -3'	(Hobel et al. 2005)	
1492r	5'- GGY TAC CTT GTT ACG ACT T -3'	(Lane 1991)	Amplification of 16S rRNA gene fragments
27f	5'- GAG TTT GAT CMT GGC TCA G -3'	(Lane 1991)	
534r	5'- ATT ACC GCG GCT GCT GG -3'	(Muyzer et al. 1993)	
342f	5'- CCT ACG GGA GGC AGC AG -3'	(Muyzer et al. 1993)	
790f	5'- GAT ACC CTG GTA GTC C -3'	(Thiel et al. 2007)	
T7	5'- T TAA TAC GAC TCA CTA TAG GG -3'	(Lloyd et al. 2004)	Amplification of the DNA-sequence inserted in the T7 operon of the pET17 vector
T7 term	5'- CTA GTT ATT GCT CAG CGG T -3'	(Lloyd et al. 2004)	
pQEf	5'- GTA TCA CGA GGC CCT TTC GTC T -3'	Designed according to QIAGEN	Amplification of the DNA-sequence inserted in the pQE30 UA vector
pQEr	5'- CAT TAC TGG ATC TAT CAA CAG GAG -3'	Designed according to QIAGEN	

Material and Methods

2.8. Sequence analyses

16S-rRNA gene comparison

The 16S-rRNA gene sequences of bacterial strains considered to possess chitinolytic capabilities were edited using ChromasPro 1.34 (Technelysium Pty. Ltd. Cologne, Germany). Closest relatives were determined by comparison to 16S rRNA genes in the National Center for Biotechnology Information (NCBI) Genbank database using the Basic Local Alignment Search Tool (BLAST) (Altschul et al. 1990). After editing, sequences were aligned by means of the FastAlign function included in the ARB software package (www.arb-home.de) and refined manually regarding the secondary structure information (Ludwig et al. 2004). Phylogenetic trees were calculated with the PhyML software (Guindon & Gascuel 2003, Guindon et al. 2005) by the Maximum Likelihood (ML) method, using the general time reversal (GTR) model and estimated proportion of invariable sites as well as Gamma distribution parameter with almost complete sequences ($\geq$1000 bp). Partial sequences (<1000 bp) were added without changing the trees topologies, using the parsimony method in ARB. Confidence limits were estimated by 500 bootstrapping replicates. The construction of the trees was partly conducted by Andrea Gärtner.

Amino acid sequence comparison

Amino acid sequences of glycoside hydrolase family 18 and 19 chitinases with the respective subfamilies were obtained from NCBI Genbank (www.ncbi.nlm.nih.gov) and the Carbohydrate Active Enzymes Online Database (CAZy) (Cantarel et al. 2009). The chitinase sequences were selected according to Karlsson and Stenlid (2009). Sequence alignments of the obtained sequences and the *Halobacterium salinarum* and *Sulfolobus tokodaii* chitinase were constructed with the Neighbour-joining method of ClustalX (Thompson et al. 1997) using the GONNET matrix. Phylogenetic trees were constructed using both the Neighbour-joining option of ClustalX as well as the Maximum-likelihood method of PROML (Phylip, version 3.6). Confidence limits were estimated by 1000 bootstrapping replicates.

Material and Methods

Sequence analysis of the archaeal *chi* genes

The physical and chemical properties of the *Sulfolobus tokodaii* and *Halobacterium salinarum* chitinase were predicted using the ProtParam tool on the ExPASy Proteomics Server (Gasteiger et al. 2003). The respective amino acid sequences of the *Sulfolobus tokodaii* chitinase and the *Halobacterium salinarum* chitinase were aligned with conserved domains of the glycoside hydrolase families 18 and 19 and whole chitinase sequences (all obtained from NCBI Genbank) using ChromasPro 1.34 (Technelysium Pty. Ltd. Cologne, Germany). To further clarify the emerging secondary structures of the *Sulfolobus tokodaii* chitinase, the amino acid sequence was submitted to the PSIPRED Protein Structure Prediction Server and analysed using the PSIPRED v3.0 program (McGuffin et al. 2000).

2.9. Cloning of archaeal chitinases

The PCR products were purified using the MinElute PCR Purification Kit (QIAGEN). The *Sulfolobus tokodaii* DNA was prepared for cloning by adding "A" overhangs with the QIAGEN A-Addition Kit (QIAGEN). The PCR product was ligated into the vector pQE-30UA (QIAGEN) with the QIAGEN UA Cloning Kit (QIAGEN) and transformed into chemically competent *E. coli* JM109 cells (Stratagen). Additionally, the plasmid was purified and transferred into the expression strain *E. coli* Bl21 cd+. The *Halobacterium salinarum* DNA and the pET17 (Novagen) vector were digested with the restriction enzymes BamH1 and Nde I (Fermentas) using the supplied Tango buffer (Fermentas). DNA and Primer were ligated using the rapid DNA ligation kit (Fermentas) and transformed into chemically competent *E. coli* BL21 cd+ cells (Stratagen). Cells were grown on LB medium (Miller 1972) containing 50 µg/ml carbenicillin and additionally 100 µg/ml chloramphenicol for *E. coli* BL21 cd+. Colonies were picked and screened for the correct insert by using the described colony PCR method.

Material and Methods

Overexpression of recombinant chitinases

A positive clone was chosen for the *Halobacterium salinarum* and for the *Sulfolobus tokodaii* approach. The clones were grown in liquid LB medium (Miller 1972) containing 50 µg/ml carbenicillin (Merck) at 37 °C, 120 RPM. For *E. coli* Bl21 cd+ 100 µg/ml chloramphenicol (Merck) were additionally added to the medium. Overexpression was induced by addition of 0.4 mM IPTG (Merck) as the cell culture reached an optical density of 1. Cells were harvested by centrifugation after 18 h of further growth.

Purification of the recombinant archaeal chitinases

The harvested *E. coli* cells were resuspended in 0.1 M citric acid (Fisher), pH 3 for the *Sulfolobus tokodaii* chitinase or in 0.1 M sodium acetate (NaAc, Fisher), pH 6 for the *Halobacterium salinarum* chitinase respectively. Buffers for handling of the *Halobacterium salinarum* protein were either amended with 0.02 M $MgCl_2$ (Merck) and 10 % (v/v) glycerol (Sigma) or with 1.5 M KCl (Sigma). The cells were ruptured with a French pressure cell at 18000 PSI (SLM Aminco, G. Heinemann). After cell lysis, the sample was centrifuged at 17700 x g for 30 min (Beckman). For *Sulfolobus tokodaii* protein the supernatant was additionally incubated at 60 °C for 1 hour, followed by a second centrifugation step at 17700 x g for 30 min. The supernatant was diluted in 0.1 MTris/HCl (Merck) buffer, pH 8.5 and applied to a HiLoad 26/10 Q Sepharose High Performance ion exchange column (GE Healthcare), equilibrated in 0.1 M Tris/HCl (Merck) buffer, pH 8.5. The respective protein was eluted by a gradient of 0 M to 2 M NaCl (Roth) in 0.1 M NaAc (Roth) buffer pH 5, with a flow rate of 8 ml/min. Active fractions were concentrated on a 30 kDa filter (Millipore). The obtained concentrate was applied to a HiLoad 16/60 Superdex 200 prep grade column (GE Healthcare), equilibrated in 0.1 M NaAc (Roth) buffer pH 5 containing 150 mM NaCl (Roth) for the *Sulfolobus tokodaii* protein and 0.1 M NaAc (Roth) buffer pH 6 containing 150 mM NaCl (Roth), $MgCl_2$ (Merck) and 10 % (v/v) glycerol (Sigma) for the *Halobacterium salinarum* enzyme. Elution was performed with a flow rate of

1 ml/min, to obtain the fraction containing the purified enzyme. Fractions were tested for chitinase activity and purity of enzyme as described below.

Molecular weight of the chitinases

The molecular weight of the respective protein was calculated according to the retention time on a HiLoad 16/60 Superdex 200 prep grade column (GE Healthcare). The column was calibrated with dextran blue (2000 kDa), ovalbumin (43 kDa), chymotrypsin (25 kDa), amylase (200 kDa), and bovine serum albumin (fraction V, 66 kDa, all from GE Healthcare).

Characterisation of the purified archaeal chitinases

The chitinase activity of the *Halobacterium salinarum* protein was determined using two different buffer systems. Both buffers consisted of 0.1 M NaAc (Roth), pH 6. In one buffer 0.02 M $MgCl_2$ (Merck) and 10 % (v/v) glycerol (Sigma) were added. In the other buffer 1.5 M KCl (Sigma) was added. For both proteins (from *Sulfolobus tokodaii* and *Halobacterium salinarum*) the chitinase activity was measured as described below (see 2.10.). For the *Sulfolobus tokodaii* protein additional parameters were measured. The pH optimum was determined by the incubation of the enzyme (0.014 mg/ml) with buffers reaching from pH 3 to pH 7 (0.1 M sodium citrate (Roth) for pH 3 and pH 4; 0.1 M NaAc (Roth) for pH 5-pH 7) in discrete steps of 1 pH at 60 °C. The temperature optimum was determined by incubation of the protein (0.017 mg/ml) at temperatures ranging from 30-80°C in discrete steps of 10°C at pH 5 (0.1 M NaAc (Roth) buffer). K_M and v_{max} values were determined by incubation of the enzyme (0.026 mg/ml) in 0.1 M NaAc (Roth) buffer pH 5 at 28°C with concentrations of colloidal chitin from 0.5-5 mg/100 µl in discrete steps of 0.5 mg.

SDS PAGE

The enzyme purity at various steps of the purification procedure and the molecular mass of the single subunit of the proteins were checked by SDS-PAGE in 12% polyacrylamide (Roth) gels followed by staining with coomassie brilliant blue R 250 (Merck) according to standard procedures (Lämmli 1970).

2.10. Chitinase activity

Protein solutions (0.1 µg- 90 µg) and respective 0.1 M buffers (100µl each) were incubated with and without 5 mg of colloidal chitin for up to 15 h at 28 °C-80 °C on a shaker. After incubation, samples and blanks were centrifuged at 13000 x g for 15 min. The release of N-acetyl-glucosamine (GlcNAc) into the supernatant was determined using the 3-methyl-2-benzothiazoline hydrazine (MBTH) method (Horn & Eijsink 2004). In this two step colorimetric process the reducing end of the chitin molecule undergoes a condensation reaction with a single MBTH molecule under alkaline conditions and elevated temperatures (step 1), followed by a second addition of a MBTH molecule under acid and oxidising conditions (step 2). The reaction yields a highly coloured product (λ_{max} at 620 nm) that can be measured photometrically (Anthon & Barrett 2002). One unit is defined as the release of 1 µmol GlcNAc per minute.

Protein retrieval from cultures of bacterial isolates and strains

Liquid cultures (100 ml) of the respective environmental strains were centrifuged for 30 min at 4618 x g. The total protein in the supernatant was precipitated by the addition of ammonium sulphate (Roth) until a final saturation of 85 % was reached (Scopes 1994) and pelleted by centrifugation for 30 min at 4618 x g. The protein pellet was rehydrated with 500 µl of deionised water (Millipore).

Protein concentration

Protein concentration was determined with the coomassie dye binding assay according to Bradford (1976), as modified by Zor (1996) with bovine serum albumin fraction V (Sigma) as standard.

3. Results

In the present work a comprehensive chitinase screening panel combining culture-based and molecular methods was established.

With the established test panel the chitin degrading capabilities of marine bacteria, either isolated from different natural habitats or selected from a culture collection were investigated and described.

In addition, the panel was also used to search for archaeal chitinase genes in the Genbank (NCBI) database and asses the chitinolytic capabilities of archaeal proteins. Thereby the first crenarchaeal chitinase gene from *Sulfolobus tokodaii* was detected, overexpressed and the respective protein was purified and characterised as a functional chitinase. Furthermore, the chitin degrading capability of the gene product of the putative chitinase gene of the euryarchaeon *Halobacterium salinarum* was demonstrated after expression in *E. coli*.

Results

3.1. Establishment of a screening panel for chitin degrading microorganisms

To establish a more comprehensive approach for the detection of chitin degraders, a three step screening approach (Fig. 3.1) was developed in this study. The screening panel consisted of the isolation and cultivation of marine microorganisms on chitin medium (Isolation of chitinolytic bacteria), the detection of the chitinolytic potential with cultivation and genetic screening of the isolated strains (detection of chitinolytic potential/activity) and the determination of the specific chitinase activity of excreted or purified enzymes (biochemical proof of chitinolytic activity).

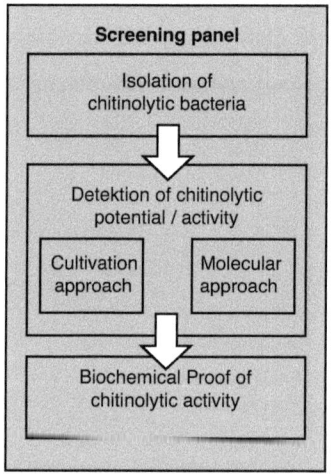

Figure 3.1: Schematic overview of the established screening panel.

3.1.1. Isolation of chitinolytic bacteria

Isolation of bacteria using solid medium

The first step in the detection of chitinolytic bacteria was the cultivation of bacteria on specific media (Fig. 3.2). For this reason a chitin medium containing purified chitin from a natural source was developed during this work (see 2.2.). A sample of *Palaemon adspersu*s was used as a source for chitinolytic bacteria (see 2.1., Tab. 2.1). The whole shrimp from the Baltic Sea was homogenised and dilution series were plated onto chitin medium (see 2.2.). 38 bacterial strains were isolated and cultured on chitin medium. All isolates were able to form colonies within 3 days when growing on chitin medium: 16 (42 %) of the strains showed the formation of clearing zones (Fig. 3.2) after incubation for up to 48 days at 28 °C (Tab. 3.1). From the 38 isolated bacterial strains only 14 were further studies, as a 100 % resemblance of the 16S-rDNA was detected in the respective strains. To exclude agar as nutritional source and to test the chitinolytic capabilities of strains growing without a clearing zone a liquid medium was designed in this study.

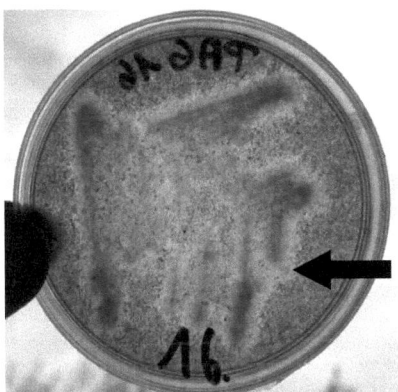

Figure 3.2: Formation of clearing zone (black arrow) around colonies of chitinolytic bacteria growing on chitin medium. Strain PAD16, after growing 48 days at 28 °C

Results

Table 3.1: List of isolates obtained from *Palaemon adspersus* homogenate, plated on chitin medium.

ID	Strain name	Occurrence of first colonies	Occurrence of clearing zone
PAD-01	*Exiguobacterium* sp.	after 3 days	yes, after 15 days
PAD-02*	*Pseudoalteromonas porphyrae*	after 3 days	yes, after 20 days
PAD-03*	*Pseudoalteromonas* sp.	after 3 days	no
PAD-04*	*Pseudoalteromonas* sp.	after 3 days	no
PAD-05*	*Pseudoalteromonas* sp.	after 3 days	no
PAD-06*	*Pseudoalteromonas* sp.	after 3 days	no
PAD-07*	*Pseudoalteromonas* sp.	after 3 days	no
PAD-08*	*Pseudoalteromonas* sp.	after 3 days	yes, after 22 days
PAD-09*	*Pseudoalteromonas* sp.	after 3 days	no
PAD-10	*Pseudoalteromonas* sp.	after 3 days	yes, after 30 days
PAD-11	*Pseudoalteromonas* sp.	after 3 days	no
PAD-12	*Pseudoalteromonas* sp.	after 3 days	yes, after 16 days
PAD-13*	*Pseudoalteromonas* sp.	after 3 days	yes, after 21 days
PAD-14*	*Pseudoalteromonas* sp.	after 3 days	no
PAD-15*	*Pseudoalteromonas* sp.	after 3 days	yes, after 18 days
PAD-16	*Vibrio* sp.	after 3 days	yes, after 48 days
PAD-17*	*Pseudoalteromonas* sp.	after 3 days	no
PAD-18*	*Pseudoalteromonas* sp.	after 3 days	no
PAD-19*	*Pseudoalteromonas* sp.	after 2 days	no
PAD-20	*Pseudoalteromonas* sp.	after 3 days	no
PAD-21*	*Pseudoalteromonas* sp.	after 3 days	no
PAD-22*	*Pseudoalteromonas* sp.	after 3 days	no
PAD-23*	*Pseudoalteromonas* sp.	after 3 days	yes, after 33 days
PAD-24*	*Aquimarina muelleri*	after 1 day	yes, after 35 days
PAD-25*	*Flavobacteriaceae* bacterium	after 1 day	no

Results

ID	Strain name	Occurrence of first colonies	Occurrence of clearing zone
PAD-26*	*Pseudoalteromonas* sp.	after 1 day	no
PAD-27	*Pseudoalteromonas* sp.	after 1 day	no
PAD-28	*Pseudoalteromonas* sp.	after 1 day	yes, after 42 days
PAD-29	*Ralstonia detusculanense*	after 2 days	yes, after 40 days
PAD-30	*Pseudoalteromonas citrea*	after 1 day	yes, after 19 days
PAD-31*	*Pseudoalteromonas* sp.	after 1 day	no
PAD-32*	*Pseudoalteromonas* sp.	after 1 day	yes, after 27 days
PAD-33	*Micrococcus luteus*	after 1 day	no
PAD-34	*Aquimarina muelleri*	after 1 day	yes, after 25 days
PAD-35*	*Pseudoalteromonas* sp.	after 1 day	yes, after 30 days
PAD-36	*Micrococcus luteus*	after 1 day	no
PAD-37	*Cellulophaga pacifica*	after 2 days	no
PAD-38*	*Cytophaga* sp.	after 3 days	no

* Strains were not used further in this study, as the 16S-rDNA sequences resembled 100 % in the respective strains.

3.1.2. Detection of chitinolytic potential / activity

Cultivation of bacteria using liquid chitin medium, cultivation approach

The liquid chitin medium contained purified chitin from a natural source as sole carbon and nitrogen source (see 2.2.). For the establishment of the liquid chitin medium, sediment samples from the eastern Mediterranean Deep Sea (see 2.1., Tab. 2.1) were homogenised and dilution series were plated on chitin medium. 67 strains were isolated and singled out by repetitive plating on chitin medium. Each isolated pure strain was transferred into liquid chitin medium. 27 of the transferred strains were able to grow in liquid chitin medium (see Tab. 3.2). As growth conditions can alter a bacterial community, solid and liquid chitin media were compared in regard to the

Results

composition of the growing phylogenetic bacterial classes. In both approaches, 9 *Bacilli* strains were growing, corresponding to 25 % on solid and 33 % in liquid medium; 6 *Actinobacteria* (class) strains were growing, corresponding to 17 % on solid and 22% in liquid medium and 1 *Flavobacteria* strain was growing, corresponding to 3 % on solid and 4 % in liquid medium. Only the number of *Gammaproteobacteria* varied between the two approaches with 20 strains (56 %) growing on the solid medium and 11 strains (41 %) growing in the liquid medium. The distribution pattern of phylogenetic classes within the cultured community changed only slightly and all phylogenetic classes that were detected in the solid medium were also detected in the liquid medium (Fig. 3.3).

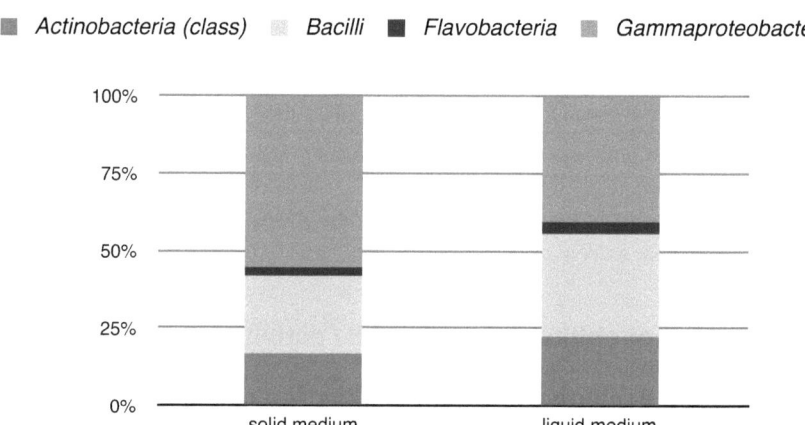

Figure 3.3: Comparison of the proportion of members of phylogenetic classes of a bacterial community isolated from eastern Mediterranean Deep Sea sediments, cultured on solid medium (left, n=36) and re-cultured in liquid medium (right, n=27).

Molecular approach

The second step during the establishment of the panel was the search for genetic fingerprints of chitinases. To accomplish this task, two degenerated primer sets designed by Hobel and co-workers (Hobel et al. 2005) were used (see 2.6., amplification of ChiA motif). The primer sets were designed to match the glycoside hydrolase subfamily 18A, the prevailing chitinase family within the bacterial clade (Metcalfe et al. 2002). A positive amplification yielded DNA fragments of 270-300 bp length. The primer systems were tested on the same 36 isolates obtained from eastern Mediterranean Deep Sea sediment samples that were used in the comparison of growth conditions (see above). 13 (36.1%) strains were positive for the primer set ChiAf1 and ChiAr1; 12 (33.3 %) strains were positive for primer set ChiAf2 and ChiAr2; 11 (30.5 %) strains were positive for both primer sets; 14 (38.9 %) strains were positive for at least one of the primer sets and 22 (61.1 %) strains were negative for both primer sets (see Tab. 3.2). In most cases strains that grew on solid or liquid chitin medium were positive for at least one primer pair. However, some strains did not possess the genetic GH18 ChiA motif but grew on chitin medium and *vice versa*.

Another molecular method called "genome mining" was used for the screening of archaeal sequences. In this approach a known bacterial chitinase amino acid sequence was used to search for similar amino acid structures in the NCBI Genbank online database with the BLAST P tool as described further on (see 3.3.).

Results

Table 3.2: Strains isolated from eastern Mediterranean Deep Sea sediment samples. Primer set 1: ChiAf1 and ChiAr1, primer set 2: ChiAf2 and ChiAr2

ID	Strain name	Growth in liquid medium	Primer set 1	Primer set 2
D02	*Stenotrophomonas maltophilia*	yes	positive	positive
D04	*Bacillus circulans*	no	negative	negative
D06	*Marinobacter flavimaris*	yes	negative	negative
D112	*Halobacillus karajiensis*	yes	negative	negative
D113	*Marinobacter salsuginis*	no	positive	positive
D114	*Pseudoalteromonas tetraodonis* strain Do-17	no	positive	positive
D117	*Leeuwenhoekiella blandensis*	yes	positive	positive
D33	*Alteromonas litorea*	yes	negative	negative
D34	*Pseudoalteromonas elyakovii*	yes	positive	positive
D35a	*Bacillus subtilis*	yes	positive	negative
D35x	*Marinobacter flavimaris*	no	positive	positive
D36	*Marinobacter flavimaris*	no	negative	negative
D37	*Pseudomonas stutzeri*	no	negative	negative
D38	*Pseudomonas stutzeri*	no	positive	positive
D39	*Alteromonas macleodii*	yes	negative	negative
D42	*Alteromonas marina* strain SW-47	no	positive	positive
D43	*Pseudomonas stutzeri* strain 4C68	no	positive	positive

Results

ID	Strain name	Growth in liquid medium	Primer set 1	Primer set 2
D44	*Micrococcus luteus*	yes	negative	negative
D45	*Alteromonas addita*	yes	negative	negative
D47	*Alteromonas addita*	yes	negative	negative
D48	*Alteromonas marina* strain SW-47	yes	positive	negative
D49	*Stenotrophomonas maltophilia*	yes	positive	positive
D50	*Pseudomonas stutzeri*	yes	negative	negative
D52	*Bacillus axarquiensis*	yes	negative	negative
D56	*Alteromonas addita*	yes	negative	negative
D81	*Streptomyces sampsonii*	yes	negative	positive
D81a	*Bacillus arsenicus*	yes	negative	negative
D92	*Streptomyces sampsonii*	yes	negative	negative
S06	*Streptomyces flavofuscus* strain NRRL B-8036	yes	positive	positive
S08	*Arthrobacter tecti*	yes	negative	negative
S08b	*Arthrobacter tecti*	yes	negative	negative
S09	*Pseudoalteromonas elyakovii* strain BSi20610	yes	negative	negative
S10	*Bacillus novalis*	yes	negative	negative
S11	*Bacillus foraminis*	yes	negative	negative
S33	*Bacillus decolorationis*	yes	negative	negative
S33x	*Bacillus firmus*	yes	negative	negative

3.1.3. Biochemical proof of chitinolytic activity

The third step within the panel was the evaluation of the chitinolytic activity of the respective enzymes. Special attention was given to excreted chitinase in the case of bacterial isolates. After incubation the liquid cultures were centrifuged and the proteins of the supernatant were recovered after precipitated. The protein fraction was investigated for chitinolytic activity in a coupled colorimetric assay that was developed in this study. Colloidal chitin, a chitin with medium to long chains, was used as substrate to overcome the problematic of false chitinase activity reads due to the use of chitin oligomers (see 1.3.). The amount of degraded chitin was determined by measuring the increase in reducing ends, occurring during chitin degradation as proxy. The reducing ends were determined by the MBTH method (see 2.10.). This method was chosen due to its accuracy and very low detection limit. After priming the system with N-acetyl-glucosamine as standard, the detection limit of the method was set to 0.02 mM of N-acetyl-glucosamine. Comparable methods (Imoto & Yagishita 1971, Garcia et al. 1993) have detection limits of 1-0.1 mM of N-acetyl-glucosamine. In addition, the MBTH method is insensitive to differing chitin chain lengths (Horn & Eijsink 2004), allowing for more accurate measurements. In addition to the establishment of this test assay the needed volumes were successfully reduced to 200 µl as final volume. This allowed the measurement to be done in a microtiterplate reader and significantly reduced the amount of enzyme needed to test for chitinase activity. The assay was tested on bacterial isolates obtained from *Carcinus maenas* samples from the Baltic Sea to further broaden the tested sources for bacterial chitin degraders. The sample was homogenised and dilution series were plated on chitin medium. A total of 8 bacterial strains were isolated and subsequently cultured in liquid chitin medium. All strains grew in liquid chitin medium. In two strains a chitinase activity was detected in the supernatant with specific activities of 16 mU and 19 mU respectively (Tab. 3.3).

Table 3.3: Bacterial strains isolated from homogenised *Carcinus maenas* sample from the Baltic Sea, with specific chitinase activity of excreted enzyme.

ID	Strain name	Growth in liquid medium	Specific chitinase activity [mU/mg]
G1	*Rhodococcus* sp.	yes	0
G2	*Shewanella* sp.	yes	0
G3	unidentified marine bacterium	yes	0
G4	*Glaciecola* sp.	yes	0
G5	unidentified marine bacterium	yes	0
G6	*Streptomyces* sp.	yes	16.2
G7	*Bacillus hwajinpoensis*	yes	0
G8	unidentified marine bacterium	yes	18.6

With this final step of the test panel the developmental phase was concluded, yielding a valid, and robust screening tool for chitinolytic bacteria. Based on the results of the developmental experiments of the three step test panel, consisting or the isolation of chitinolytic bacteria, the detection of chitinolytic potential/activity and the biochemical proof of chitinolytic activity, it was decided to consider bacterial strains isolated on chitin medium that either possess the glycoside hydrolase family 18A gene (GH18 ChiA) motif or grew in liquid chitin medium as chitinolytic bacteria. The specific activity of excreted chitinases can be validly determined in the third step of the panel. The respective parts of the screening panel were also used for detection and evaluation of archaeal chitinases (see 3.3.).

3.2. Screening of natural occurring bacteria and strain collections for chitinolytic activity

During this study, 145 bacterial strains from different marine habitats (Tab. A1, Appendix) were screened for their chitin degrading capability with the developed screening panel. The strains were isolated with chitin medium during sampling campaigns to the Baltic and Mediterranean Sea and randomly chosen from the Actinomycetes- and Bryozoan-derived bacterial section of the KiWiZ strain collection (see 2.1. and 3.1.).

In total, 133 (90.7 %) of the screened strains possessed chitin degrading potential (Fig. 3.4). 102 (70.3 %) strains grew in liquid chitin medium. 69 (47.6 %) strains possessed the GH18 ChiA motif. 38 strains (26.2 %) possessed the GH18 ChiA motif and grew under laboratory conditions on chitin as sole carbon source. 31 strains (21.4 %) possessed the genetic GH18 ChiA motif, but did not grow under laboratory conditions with chitin as sole carbon source, whereas 64 strains (44.1 %) grew on liquid chitin medium but did not possess the genetic GH18 ChiA motif (Tab. A1, Appendix).

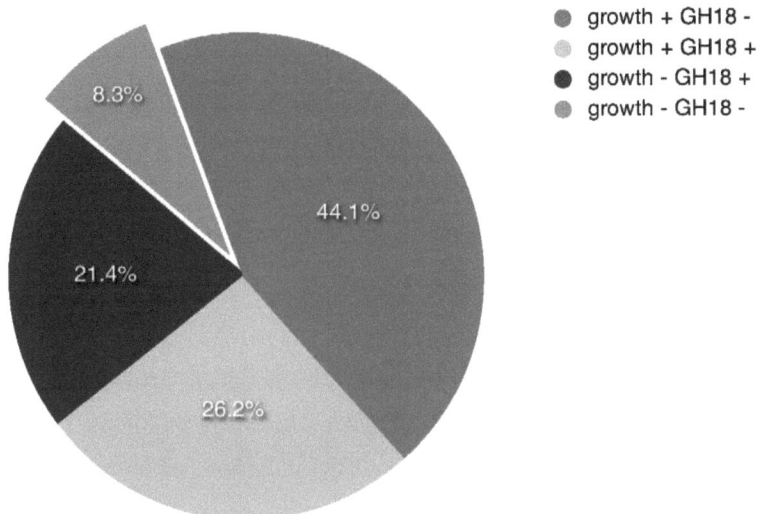

Figure 3.4: Distribution of growing and non growing strains with chitin as sole carbon source, as well as the strains possessing or not possessing the genetic GH18 ChiA motif within the 145 isolates of this study. Grey: strains were growing in liquid chitin medium but no GH18 ChiA motif was detectable. Light grey: strains were growing in liquid chitin medium and the GH18 ChiA motif was detected. Dark grey: strains did not grow in liquid chitin medium but the GH18 ChiA motif was detected. Middle grey: the strains did neither grow in liquid chitin medium nor was the GH18 ChiA motif detectable.

Comparison of marine habitats

The investigated marine habitats were eastern Mediterranean Deep Sea sediments, western Mediterranean Deep Sea sediments and carapaces of the Baltic Sea shrimp *Palaemon adspersus*. Bacteria from these habitats were isolated on chitin medium and screened for their chitin degrading capability. The pure cultures were additionally sequenced to obtain 16S rDNA sequences for a phylogenetic analysis.

From the Mediterranean, 53 strains were screened for chitinolytic activity in this study. The strains were classified as follows.

Eastern Mediterranean Deep Sea sediment (36 strains): 33 (91.7 %) strains possessed chitinolytic potential. 6 (16.7 %) strains possessed the genetic capability, but did not grow on chitin as sole carbon source under laboratory conditions. 19 (52.8 %) strains did not possess the GH18 CHiA motif but did grow on chitin as sole carbon source under laboratory conditions. 8 (22.2 %) strains possessed the genetic capability and did grow on chitin as sole carbon source under laboratory condition. The bacterial chitin degrading community was dominated by *Gammaproteobacteria* (17 isolates, 51.1 %), followed by *Bacilli* (9 isolates, 27.3 %), *Actinobacteria* (6 isolates, 18.2 %) and *Flavobacteria* (1 isolate, 3 % Fig. 3.5).

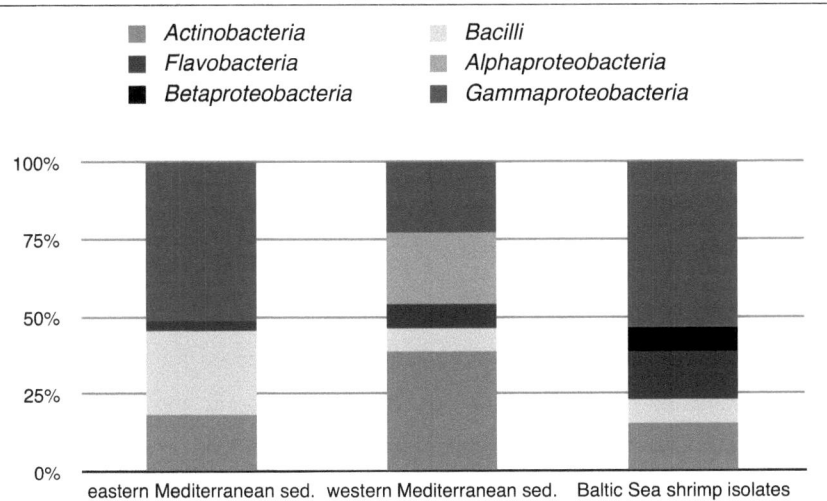

Figure 3.5: Comparison of the phylogenetic classes of the bacterial communities isolated from different habitats. Left: eastern Mediterranean sediment n=36. Middle: western Mediterranean sediment n=18. Right: shrimp carapaces n=13.

Western Mediterranean Deep Sea sediments (18 strains): 13 (72.2 %) strains possessed chitinolytic potential. 1 (5.6 %) strain possessed the genetic capability but did not grow on chitin as sole carbon source under laboratory conditions. 7 (38.9 %) strains did not possess the GH18 CHiA motif but did grow on chitin as sole carbon source under laboratory conditions. 5 (27.8 %) strains possessed the genetic capability and did grow on chitin as sole carbon source under laboratory conditions. The bacterial chitin degrading community was dominated by *Actinobacteria* (5 isolates, 38.5 %), followed by *Alphaproteobacteria* (3 isolates, 23.1 %), *Gammaproteobacteria* (3 isolates, 23.1 %), *Bacilli* (1 isolate, 7.7 %), and *Flavobacteria* (1 isolate, 7.7 % Fig. 3.5).

From a homogenised carapace of the shrimp *Palemon adspersus* 14 of the isolated bacterial strains were screened for their chitinolytic potential. 13 (92.9 %) strains possessed chitinolytic potential. 2 (14.3 %) strains possessed the genetic capability but did not grow on chitin as sole carbon source under laboratory conditions. 6 (42.9 %) strains did not possess the GH18 CHiA motif but did grow on chitin as sole

Results

carbon source under laboratory conditions. 5 (35.7 %) strains possessed the genetic capability and did grow on chitin as sole carbon source under laboratory conditions. The bacterial chitin degrading community was dominated by *Gammaproteobacteria* (7 isolates, 53.8 %), followed by *Actinobacteria* (2 isolates, 15.4 %), *Flavobacteria* (2 isolates, 15.4 %), *Betaproteobacteria* (1 isolate, 7.7 %) and *Bacilli* (1 isolate, 7.7 % Fig. 3.5).

Screening of the KiWiZ strain collection

Two sections of the KiWiZ strain collection were screened for chitinolytic active bacterial strains. 45 bacterial strains isolated from bryozoans and 31 *Actinomycetes* were randomly chosen, cultivated under laboratory conditions on chitin as sole carbon and nitrogen source and screened for the presence of the glycoside hydrolase family 18 A motif.

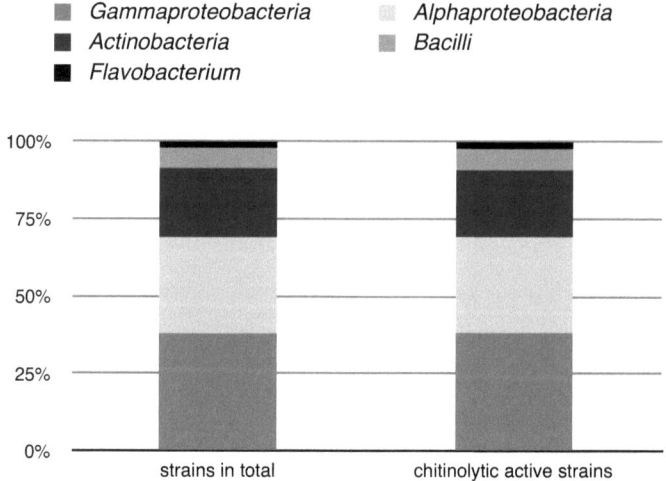

Figure 3.6: Comparison of analysed strains (strains, left) and strains with chitinolytic activity (active strains, right) in *per cent* of the bryozoan-derived bacteria.

Bryozoan-derived bacterial strains: From the 45 selected strains 42 (93.3 %) strains possessed chitinolytic potential. 18 (40 %) strains possessed the genetic capability but

Results

did not grow on chitin as sole carbon source under laboratory conditions. 19 (42.2 %) strains did not possess the GH18 ChiA motif but did grow on chitin as sole carbon source under laboratory conditions. 4 (8.9 %) strains possessed the genetic capability and did grow on chitin as sole carbon source under laboratory conditions. The screened strains consisted of 17 (37.8 % of all screened strains) *Gammaproteobacteria* with 16 (38.1 % of all chitinolytic active screened strains) chitinolytic active and 1 (2.2 % of all screened strains) non active strain. 14 (31.1 % of all chitinolytic active screened strains) *Alphaproteobacteria* with 13 (31 % of all chitinolytic active screened strains) chitinolytic active strains and 1 (2.2 % of all screened strains) non active strain. 10 (22.2 % of all screened strains) *Actinobacteria* with 9 (21.4 % of all chitinolytic active screened strains) chitinolytic active strains and 1 (2.2 % of all screened strains) non active strain. 3 (6.6 % of all screened strains, 7.1 % of all chitinolytic active screened strains) *Bacilli* (all chitinolytic active) and 1 (2.2 % of all screened strains, 2.4 % of all chitinolytic active screened strains) *Flavobacterium*, also chitinolytic active (Fig. 3.6).

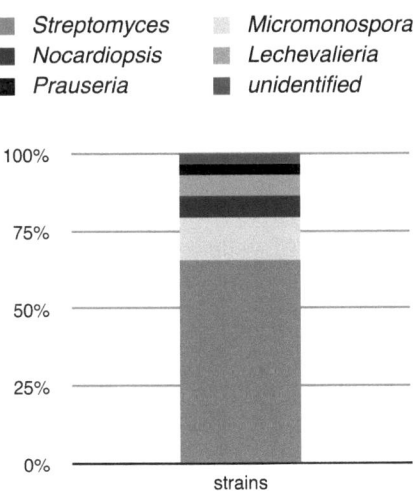

Figure 3.7: Phylogentic composition of analysed *Actinomycestes* on genus level (all active) of the KiWiZ strain collection in *per cent*.

Results

Actinomycetes: All of the 29 selected strains possessed chitinolytic potential. 2 (6.9 %) strains possessed the genetic capability but did not grow on chitin as sole carbon source under laboratory conditions. 12 (41.4 %) strains did not possess the GH18 CHiA motif but did grow on chitin as sole carbon source under laboratory conditions. 15 (51.7 %) strains possessed the genetic capability and did grow on chitin as sole carbon source under laboratory conditions The screened isolates consisted of 5 genera and one unidentified strain. 19 (65.5 %) belonged to the genus *Streptomyces*. 4 (13.8 %) belonged to the genus *Micromonaspora*. 2 (6.9 %) belonged to the genus *Nocardiopsis*. 2 (6.9 %) belonged to the genus *Lechevalieria*. 1 (3.4 %) belonged to the genus *Prauseria*. 1 (3.4 %) belonged to an unknown genus (Fig. 3.7).

Excreted chitinases

All strains that were capable to grow within this study on chitin as sole carbon and nitrogen source (102 strains) were screened for their ability to excrete a chitinolytic protein into the growth medium. In the growth broth of 12 (11.8 %) strains chitinases were detected with specific activities ranging from 0.01 to 0.42 mU/mg (Tab 3.4). Isolates excreting detectable chitinases into the growth medium belonged to the class of the *Firmicutes* (2 strains, corresponding to 16.7 %), the *Gammaproteobacteria* (3 strains, corresponding to 25 %) and to the *Actinobacteria* (7 strains, corresponding to 58.3 %). Within the *Actinobacteria* most strains excreting an active detectable chitinase belonged to the genus *Streptomyces* with 5 strains, corresponding to 71.4 % of the class *Actinobacteria* and 41 % of all detected chitinase excreting strains.

Table 3.4: Bacterial strains with detected chitinolytic protein excreted into the cultivation medium.

ID	Strain Name	Origin	Specific activity [mU/mg]
	Firmicutes		
PAD1	*Exiguobacterium* sp.	Baltic Sea	0.01
S11	*Bacillus foraminis*	Mediterranean Deep Sea	0.02
	Actinobacteria		
D81	*Streptomyces sampsonii*	Mediterranean Deep Sea	0.03
BB50a	*Streptomyces* sp.	Baltic Sea	0.04
HB217	unidentified *Actinomycetes*	Baltic Sea	0.04
i62	*Micrococcus indicus*	Mediterranean Deep Sea	0.10
HB346	*Streptomyces roseoflavus*	Baltic Sea	0.11
BB11a	*Streptomyces* sp.	Baltic Sea	0.25
HB122	*Streptomyces griseus*	Baltic Sea	0.42
	Gammaproteobacteria		
B157	*Shewanella* sp.	Baltic Sea	0.12
PAD27	*Pseudoalteromonas* sp.	Baltic Sea	0.12
PAD16	*Vibrio* sp.	Baltic Sea	0.39

Results

Phylogenetic analysis of potential chitin degrading isolates

To elucidate potential patterns within the phylogenetic distribution of chitinolytic strains, 95 (Tab. A2, Appendix) high quality 16 S-rDNA sequences with the necessary length were obtained to conduct a phylogenetic analysis. This sequences were obtained from the potentially chitinolytic active bacteria isolated during the sampling campaigns in the Mediterranean Deep Sea and to the Baltic Sea, as well as from the strains chosen from the bryozoan-derived section of the KiWiZ strain collection. The phylogenetic analysis (Fig. 3.9 A-C) showed a broad distribution of chitinolytic microorganisms in the bacterial domain of life. The different sequences were attributed to the *Alphaproteobacteria* (16 strains, corresponding to 16.84 %), the *Gammaproteobacteria* (42 strains, corresponding to 44.2 %), the *Flavobacteria* (4 strains, corresponding to 4.2 %), the *Bacilli* (14 strains, corresponding to 14.7 %) and the *Actinobacteria* (19 strains, corresponding to 20 %). Most strains were attributed to the *Gammaproteobacteria*. Strains from the different habitats were found in all the phylogenetic classes, except for the *Alphaproteobacteria*, where no isolates from shrimp carapaces were found (Fig. 3.8). Clusters attributed to different habitats were found in the *Gammaproteobacteria* and the *Bacilli*. Within the *Gammaproteobacteria* (Fig. 3.9 A) a Baltic Sea bryozoan cluster (cluster I) was detected, with *Halomonas* and *Shewanella* species. In addition, a Mediterranean Sea cluster (cluster II) was found containing only *Alteromonas* species from Mediterranean sediment and bryozoan samples. In addition, two mixed clusters were detected. The first cluster (cluster III) contained only *Pseudoalteromonas* species from all habitats and regions. The second cluster (cluster IV) contained *Pseudomonas* species from Baltic Sea bryozoans and Mediterranean sediment samples. Within the *Bacilli* (Fig. 3.9 C) a large mixed cluster (cluster V) was detected containing only *Bacillus* species from Mediterranean sediment and Baltic Sea bryozoans.

Results

With this phylogenetic analysis a deeper insight into the distribution of bacterial chitin degraders on the species level was made possible and will be debated in the discussion.

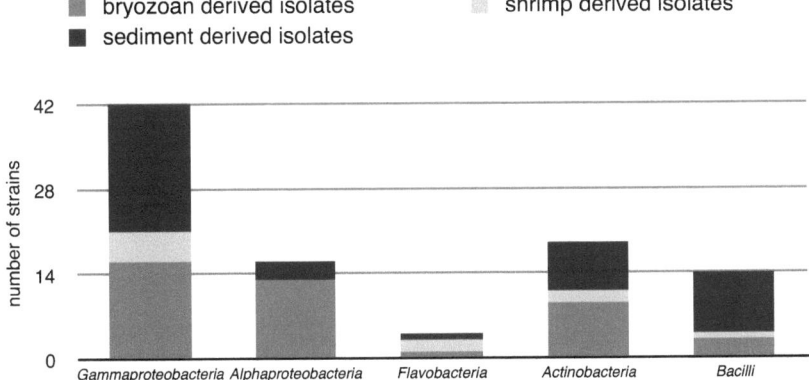

Figure 3.8: Distribution of bacterial isolates with chitinolytic potential from different marine habitats amongst phylogenetic classes.

Results

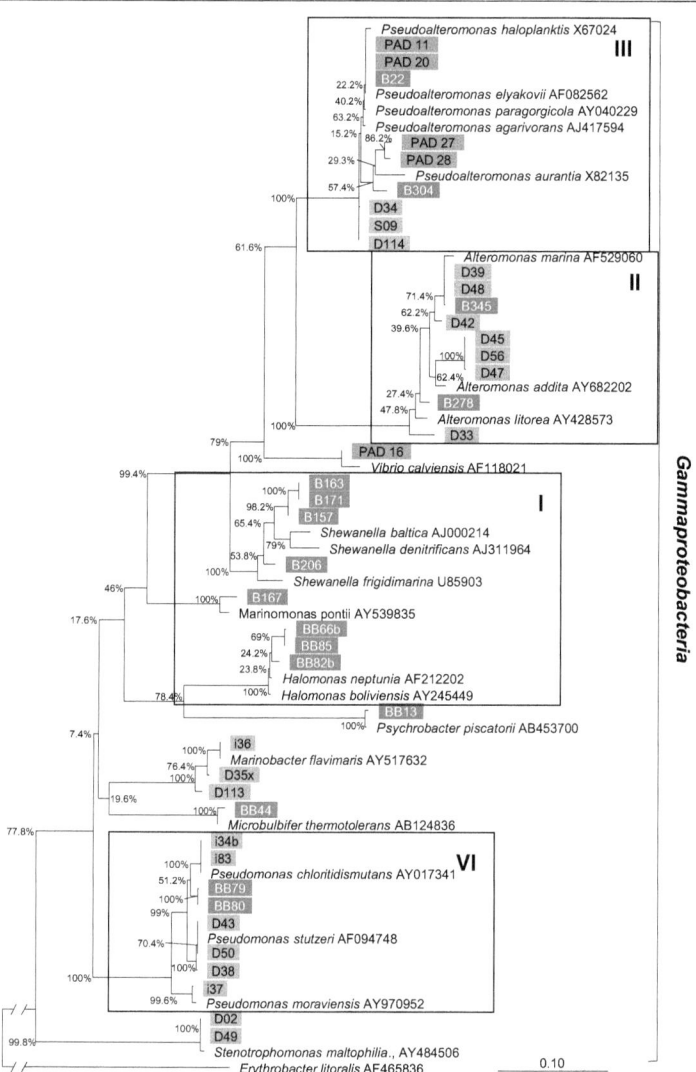

Figure 3.9 A: Phylogenetic 16S-rDNA alignment of potential chitinolytic *Gammaproteobacteria*. Prefixes were chosen randomly. Highlighted in grey with the prefix "B" or "BB": Strains isolated from various bryozoans. Highlighted in grey with the prefix "D", "i" or "S": Strains isolated from Mediterranean deep sea sediment. Highlighted in grey with the prefix "PAD": Strains isolated from Baltic Sea crab carapaces. Boxes represent the different clusters. Bootstrap values (500 replicates) are given in *per cent* at the nodes.

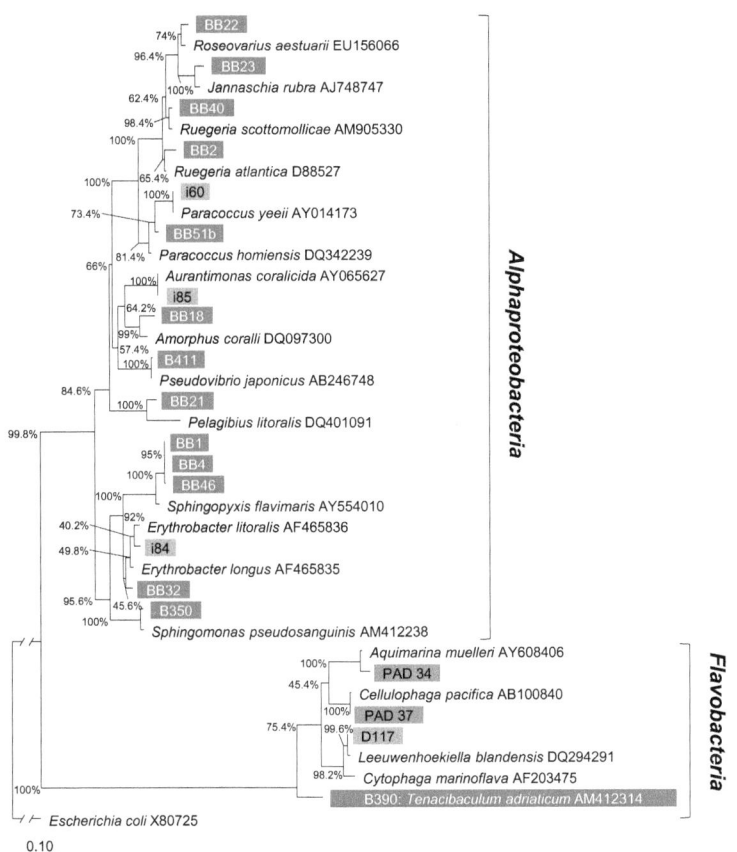

Figure 3.9 B: Phylogenetic 16S-rDNA alignment of potential chitinolytic *Alphaproteobacteria* and *Flavobacteria*. Prefixes were chosen randomly. Highlighted in grey with the prefix "B" or "BB": Strains isolated from various bryozoans. Highlighted in grey with the prefix "D" or "i": Strains isolated from Mediterranean deep sea sediment. Highlighted in grey with the prefix "PAD": Strains isolated from Baltic Sea crab carapaces. Bootstrap values (500 replicates) are given in *per cent* at the nodes.

Results

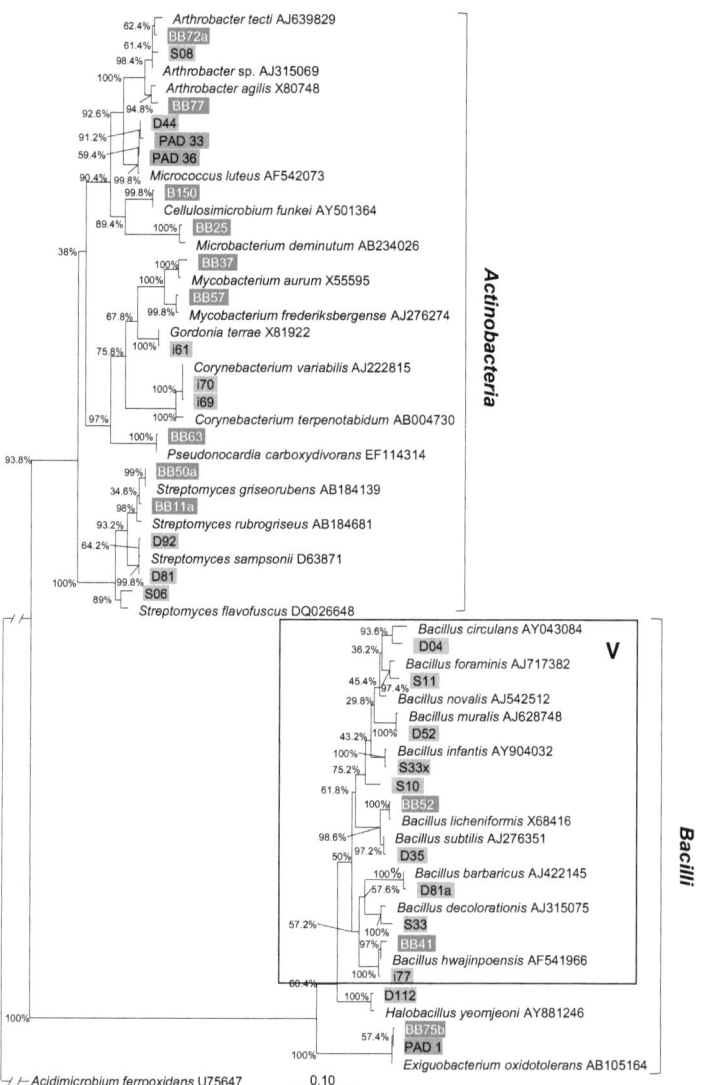

Figure 3.9 C: Phylogenetic 16S-rDNA alignment of potential chitinolytic *Actionobacteria* and *Bacilli*. Prefixes were chosen randomly. Highlighted in grey with the prefix "B" or "BB": Strains isolated from various bryozoans. Highlighted in grey with the prefix "D", "i" or "S": Strains isolated from Mediterranean deep sea sediment. Highlighted in grey with the prefix "PAD": Strains isolated from Baltic Sea crab carapaces. Box represents the detected cluster. Bootstrap values (500 replicates) are given in *per cent* at the nodes.

3.3. Detection, isolation and characterisation of archaeal chitinases

In this part of the study, the molecular screening and biochemical evaluation of the chitinase activity of enzymes established in the screening panel was used to find and characterise archaeal chitinases. Utilising genome mining (see 3.1.) with the known chitinase sequence of *Streptomyces olivaceoviridis* as template (see 2.3.), two archaeal chitinase sequences were detected: One in the genome of the crenarchaeon *Sulfolobus tokodaii* and one in the genome of the archaeon *Halobacterium salinarum*. The proposed chitinase sequences were amplified with specific primers designed in the course of this work, cloned into the expression systems (*E. coli*) and overexpressed. The resulting proteins were purified and tested for chitinase activity.

3.3.1. A novel crenarchaeal chitinase from *Sulfolobus tokodaii*

Identification of a *chi* gene from *Sulfolobus tokodaii*

Based on amino acid sequence similarity (see 2.3.), an open reading frame (ORF) was identified by BLAST P search in the completely sequenced genome of *Sulfolobus tokodaii*. This ORF, BAB65950, was predicted to encode for a "709 amino acid long hypothetical protein". The ORF contained 2130 bp coding for a polypeptide of 709 amino acids with a calculated molecular mass of 77.7 kDa. The protein had a predicted pI of 8.32. The G + C content of the ORF was 33 %. The coding sequence started with ATG and stopped with TAA.

Results

Sequence analysis of *Sulfolobus tokodaii* ORF BAB65950

The primary structure of the ORF BAB65950 was subjected to a BLAST search and analysed, using an alignment (Fig. 3.10), in order to identify the most similar enzyme as well as the most similar three-dimensional (3D)-structure-determined enzyme: The *Sulfolobus tokodaii chi* sequence could be neither matched exactly with the highly conserved DXDXE signature motif of the GH18 family (Tsuji et al. 2010), nor with the highly conserved [FHY]-G-R-G-[AP]- X-Q-[IL]-[ST]-[FHYW]-[HN]-[FY]-NY motif of the GH19 family (Huet et al. 2008).

At the N-terminal end of the *Sulfolobus tokodaii* protein, a broad-complex, tramtrack and bric a brac (BTB) domain (Bardwell & Treisman 1994) was found. Adjacent to the BTB domain, a chitin/cellulose binding domain (ChtBD3) (Brun et al. 1997) with the conserved residues Trp220 and Tyr237 was identified. Additionally, in the middle of the sequence, a fibronectin type III domain was detected (Toratani et al. 2006).

Results

Figure 3.10: Caption see next page.

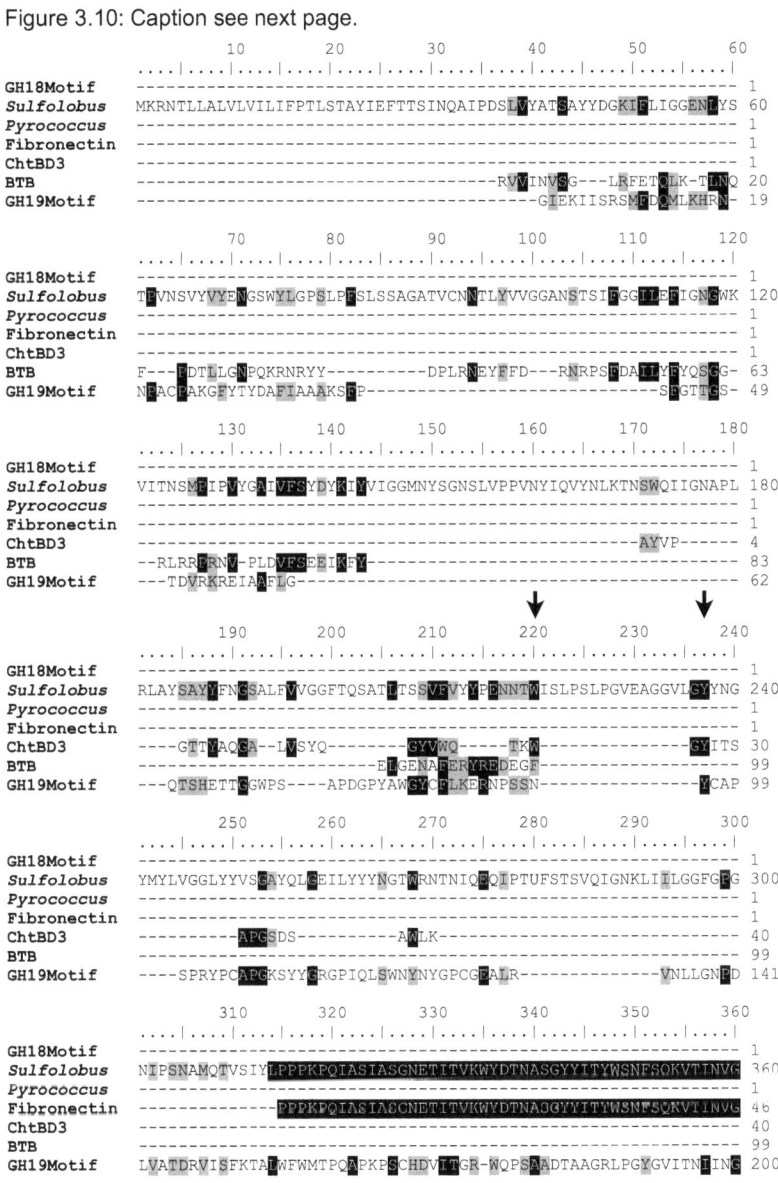

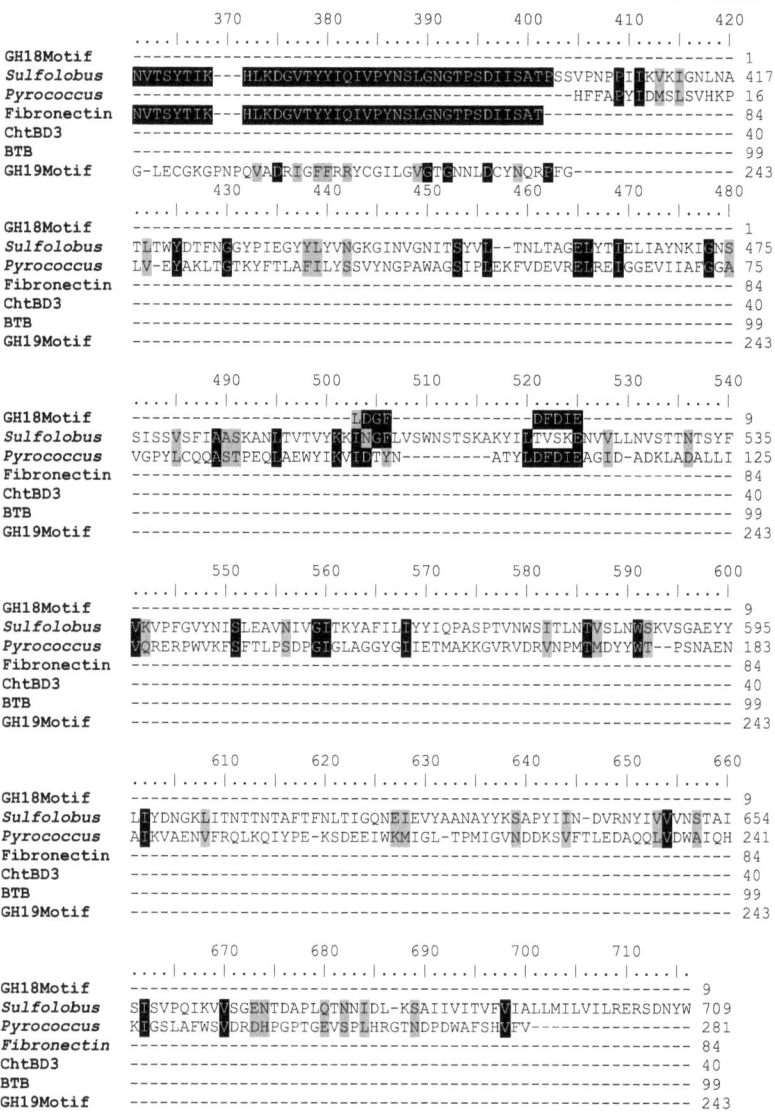

Figure 3.10: Alignment of protein sequence similarity of proposed *Sulfolobus tokodaii* chitinase with chitinase core motifs. Black boxes denote conserved amino acids; grey boxes denote similar amino acids. GH18 Motif: Consensus GH18 core motif with signal sequence and conserved glutamic acid residue. *Sulfolobus*: Amino acid sequence of ORF BAB65950 from *Sulfolobus tokodaii* (NCBI accession number: NC 003106). *Pyrococcus*: Amino acid

Results

Figure 3.10 caption continued: - sequence of the *Pyrococcus furiosus* chitinase (NCBI accession number: NC 003413). Fibronectin: Amino acid sequence of conserved fibronectin domain (NCBI accession number Q973G3 region 315 to 398). ChtBD3: Amino acid sequence of conserved chitin binding domain (NCBI accession number ABI31434.1, region 453 to 492) with conserved binding residues Trp220 and Tyr237 (black arrows). BTB: Amino acid sequence of a broad-complex, tramtrack and bric a brac domain (NCBI accession number: 1EOE_A). GH19 Motif: GH19 motif of *Carcia papaya* (NCBI accession number: 3CQL_A).

Cloning and overexpression of the proposed *chi* gene from *Sulfolobus tokodaii*

The *chi* gene was amplified by PCR and cloned into the vector pQE-30UA (QIAGEN). The recombinant plasmid was used to transform *E. coli* Bl21 cd+ and expression was induced by IPTG. The overexpressed polypeptide was examined by SDS-PAGE, showing a single band with a molecular mass of approximately 77 kDa. This result was consistent with the molecular mass predicted from the nucleotide sequences (77.7 kDa). The recombinant protein was purified by heat and acid precipitation, followed by a cationic exchange column and a gelfiltration step. According to the retention time the native molecular mass was determined to be 130 kDa, indicating a homodimer (alpha 2).

Kinetic characterisation of the *Sulfolobus tokodaii* enzyme

Utilising the MBTH method a specific activity of 75 mU/mg could be detected in the purified protein and therefore, a functional chitinase was shown to be encoded by the detected gene. The chitinase K_M value was determined to be 65.9 mg of colloidal chitin. Its optimal activity was measured at 70 °C and pH 2.5 (Fig. 3.11).

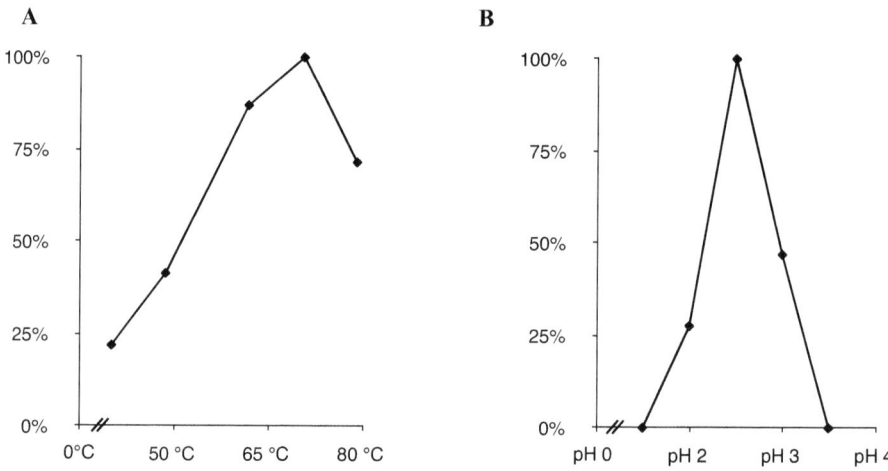

Figure 3.11: Relative activity of the *Sulfolobus tokodaii* chitinase. A: At different temperatures. 100% correspond to 1.81 mU/mg. The assays were conducted at pH 3. B: At different pH values. 100% correspond to 0.42 mU/mg. The assays were conducted at 60 °C

3.3.2. Expression of the euryarchaeal chitinase from *Halobacterium salinarum*

Identification of the *chi* gene from *Halobacterium salinarum*

Based on amino acid sequence similarity, an open reading frame (ORF) was identified by a BLAST P search in the completely sequenced genome of *Halobacterium salinarum* R1 (Pfeiffer et al. 2008). This ORF CAP13543 was predicted to encode for a chitinase. The ORF contained 1641 bp coding for a polypeptide of 546 amino acids with a calculated molecular mass of 60.22 kDa. The protein had a predicted pI of 4.08. The G + C content of the ORF was 65.6 %. The coding sequence started with ATG and stopped with TAA.

Sequence analysis of *Halobacterium salinarum*

The primary structure of the ORF CAP13543 was subjected to a BLAST search and analysed, using an alignment (Fig. 3.12), in order to identify the most similar enzyme as well as the most similar three-dimensional (3D)-structure-determined enzyme: Three conserved domains were detected. At the N-terminal side a chitin/cellulose binding domain (ChtBD3) (Brun et al. 1997) with the respective conserved residues Trp35 and Tyr41, crucial for the formation of the substrate binding cleft were detected. Adjacent to the chitin binding domain a polycystic kidney diseases I domain (PKD) (Orikoshi et al. 2005) was found. This domain is similar to other cell-surface modules, with an IG-like fold. The domain probably functions as a ligand binding site in protein-protein or protein-carbohydrate interactions (Orikoshi et al. 2005) and is also found in microbial chitinases. The C-terminal side of the protein harbours the GH 18 motif with its highly conserved DXDXE motif (Tsuji et al. 2010).

Results

Cloning and overexpression of the chitinase from *Halobacterium salinarum*

The *chi* gene was amplified by PCR and cloned into the vector pET17b (Novagen). The recombinant plasmid was used to transform *E. coli* Bl21 cd+ and induced by IPTG. The overexpressed polypeptide was examined by SDS-PAGE, showing a single band with a molecular mass of approximately 60 kDa. This result was consistent with the molecular mass predicted from the nucleotide sequences (60.22 kDa). The recombinant protein was purified by a gelfiltration step. According to the retention time the native molecular mass was determined to be 330 kDa, indicating a hexameric protein.

Kinetic characterisation of the *Halobacterium salinarum* enzyme

Utilising the MBTH method the non-purified chitinase showed a specific activity of 0.7 mU/mg when incubated with 1.5 M KCl. The specific activity increased five fold when non purified protein was incubated with 0.02 M $MgCl_2$ and 10% (v/v) glycerol to 3.3 mU/mg. A specific activity of the purified protein of 38 mU/mg was detected at pH 6 (0.1M NaAc) at 28 °C in the presence of 0.02 M $MgCl_2$ and 10% (v/v) glycerol. The enzyme was not further characterised as the main focus was the expression of a functional enzyme in *E. coli*.

Results

Figure 3.12:Caption see next page.

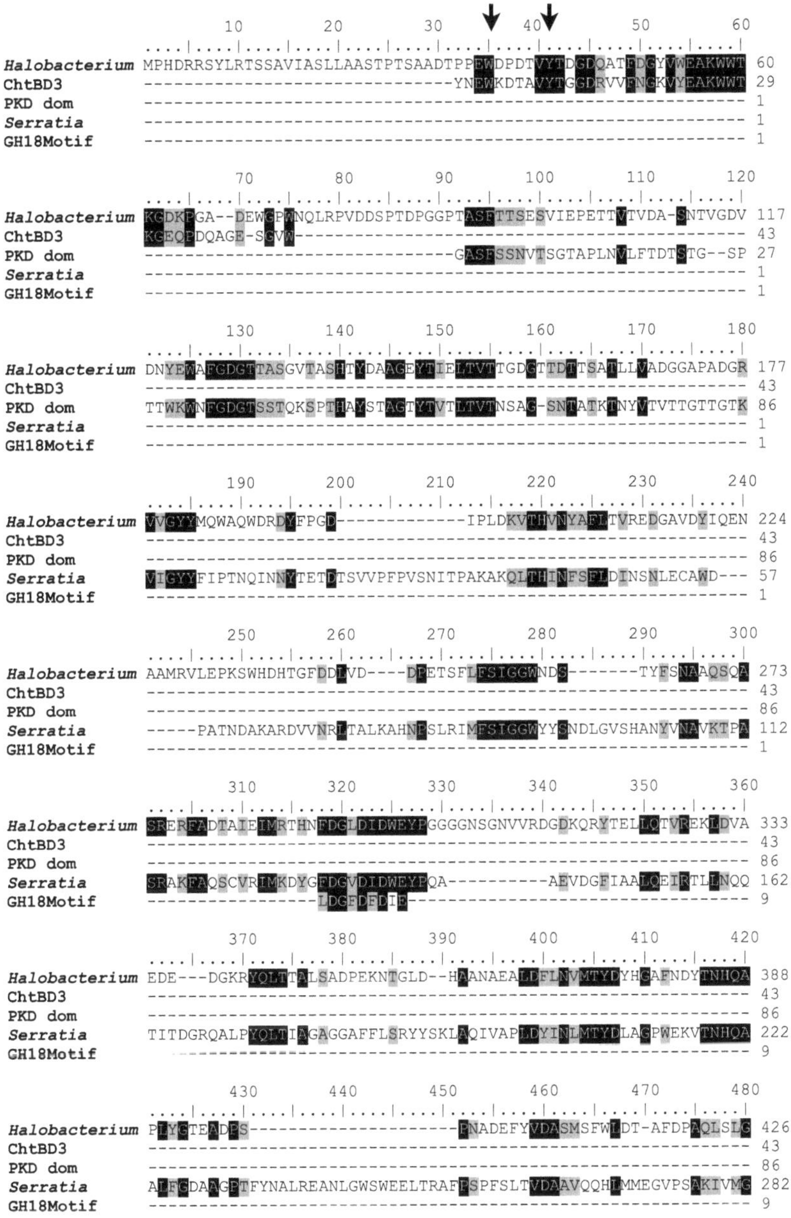

```
                        10        20        30        40        50        60
                ....|....|....|....|....|....|....|....|....|....|....|....|
Halobacterium   MPHDRRSYLRTSSAVIASLLAASTPTSAADTPPEWDPDTVYTDSDQATFDGYVWEAKWWT  60
ChtBD3          ----------------------------YNEWKDTAVYTGGDRVVFNGKVMEAKWWT   29
PKD dom         ------------------------------------------------------------ 1
Serratia        ------------------------------------------------------------ 1
GH18Motif       ------------------------------------------------------------ 1

                        70        80        90       100       110       120
                ....|....|....|....|....|....|....|....|....|....|....|....|
Halobacterium   KGDKFGA--DEWGPANQLRPVDDSPTDPGGPTASFTTSESVIEPETTVTVDA-SNTVGDV  117
ChtBD3          KGEQEDQAGE-SGVN--------------------------------------------  43
PKD dom         -------------------------------GASFSSNVTSGTAPLNVLFTDTSTG--SP 27
Serratia        ------------------------------------------------------------ 1
GH18Motif       ------------------------------------------------------------ 1

                       130       140       150       160       170       180
                ....|....|....|....|....|....|....|....|....|....|....|....|
Halobacterium   DNYEWAFGDGTTASGVTASHTYDAAGEYTIELTVTTGDGTTDTTSATLLVADGGAPADGR  177
ChtBD3          -----------------------------------------------------------  43
PKD dom         TTWKANFGDGTSSTQKSPTHAYSTAGTYTVTLTVTNSAG-SNWATKTNYVTVTTGTTGTK  86
Serratia        ------------------------------------------------------------ 1
GH18Motif       ------------------------------------------------------------ 1

                       190       200       210       220       230       240
                ....|....|....|....|....|....|....|....|....|....|....|....|
Halobacterium   VVGYYMQWAQWDRDYFPGE-------------IPLDKVTHVNYAFITVREDGAVDYIQEN 224
ChtBD3          -----------------------------------------------------------  43
PKD dom         -----------------------------------------------------------  86
Serratia        VIGYYFIPTNQINNYTETFTSVVPFPVSNITPAKAKQLTHINFSFLDINSNLECAWD---  57
GH18Motif       ------------------------------------------------------------ 1

                       250       260       270       280       290       300
                ....|....|....|....|....|....|....|....|....|....|....|....|
Halobacterium   AAMRVLEPKSWHDHTGFDDLVD----DPETSFLFSTGGWNDS-------TYFSNAAQSQA 273
ChtBD3          -----------------------------------------------------------  43
PKD dom         -----------------------------------------------------------  86
Serratia        -----PATNDAKARDVVNRLTALKAHNESLRIMFSIGGWYYSNDLGVSHANYVNAVKTPA 112
GH18Motif       ------------------------------------------------------------ 1

                       310       320       330       340       350       360
                ....|....|....|....|....|....|....|....|....|....|....|....|
Halobacterium   SRERFADTAIEIMRTHNFDGLDIDWEYPGGGGNSGNVVRDGDKQRYTELLQTVREKLDVA 333
ChtBD3          -----------------------------------------------------------  43
PKD dom         -----------------------------------------------------------  86
Serratia        SRAKFAQSCVRIMKDYGFDGVDIDWEYPQA----------AEVDGFIAALQEIRTLLNQQ 162
GH18Motif       ----------------LDGFDEDLIF---------------------------------  9

                       370       380       390       400       410       420
                ....|....|....|....|....|....|....|....|....|....|....|....|
Halobacterium   EDE---DGKRYQLTIALSADPEKNTGLD--HAANAEALLFLNVETYDYHGAFNDYTNHQA 388
ChtBD3          -----------------------------------------------------------  43
PKD dom         -----------------------------------------------------------  86
Serratia        TITDGRQALPYQLTIAGAGGAFFLSRYYSKLAQIVAPLLYINLMTYDLAGPWEKVTNHQA 222
GH18Motif       -----------------------------------------------------------  9

                       430       440       450       460       470       480
                ....|....|....|....|....|....|....|....|....|....|....|....|
Halobacterium   PLYGTEADES-------------------FNADEFYVDASMSFWLDT-AFDPRQLSLG  426
ChtBD3          ----------------------------------------------------------  43
PKD dom         ----------------------------------------------------------  86
Serratia        ALFGDAAGETFYNALREANLGWSWEELTRAFPSPFSLTVDAAVQQHLMMEGVPSAKIVMG 282
GH18Motif       ----------------------------------------------------------   9
```

67

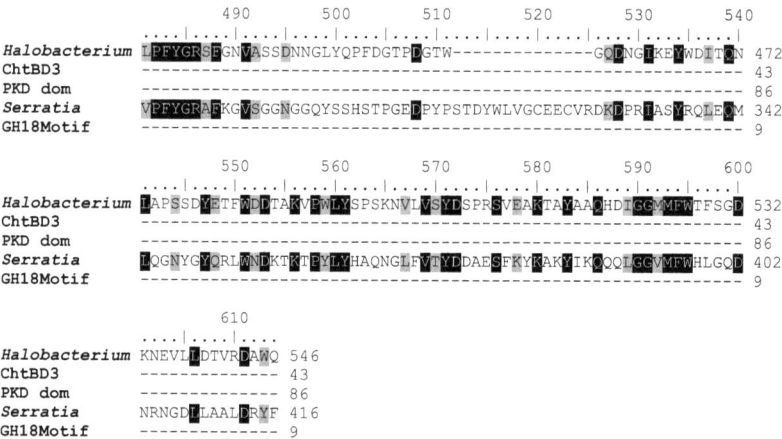

Figure 3.12: Alignment of protein sequence of the *Halobacterium salinarum* chitinase with chitinase core motifs. Black boxes denote conserved amino acids; grey boxes denote similar amino acids. *Halobacterium*: Amino acid sequence of ORF CAP13543 from *Halobacterium salinarum* (NCBI accession number: AM774415). ChtBD3: Chitin binding domain 3 from *Bacillus subtilis* (GI: 3193265) with conserved binding residues Trp35 and Tyr41 (black arrows). PKD dom: Surface layer protein 3 from *Methanosarcina mazei* (GI: 74486927), belonging to the PKD domain group. *Serratia*: Amino acid sequence of the *Serratia marcescens* chitinase B (GI: 14719596). GH18 Motif: Consensus GH18 core motif with signal sequence and conserved glutamic acid residue (Tsuji et al. 2010).

3.3.3. Phylogenetic comparison of chitinases from different domains of life

To further clarify the position of the detected archaeal chitinases within the glycoside hydrolase families 18 and 19, a phylogenetic protein tree was constructed. In total, 70 sequences from GH family 18 (n=50) and 19 (n=20) were obtained from NCBI (Tab. A3, Appendix) and aligned with ClustalX. Three distinct subfamilies GH18A, GH18B and GH18C were detected (Karlsson & Stenlid 2009). All sequences of the GH19 family formed a single, distinct group (Fig. 3.13). The euryarchaeal sequences, including the chitinase sequence from *Halobacterium salinarum* were found within the GH18 clade in the subfamilies A and C (Fig. 3.13). The crenarchaeal chitinase sequence of *Sulfolobus tokodaii* clustered into close proximity to the glycoside hydrolase family 18.

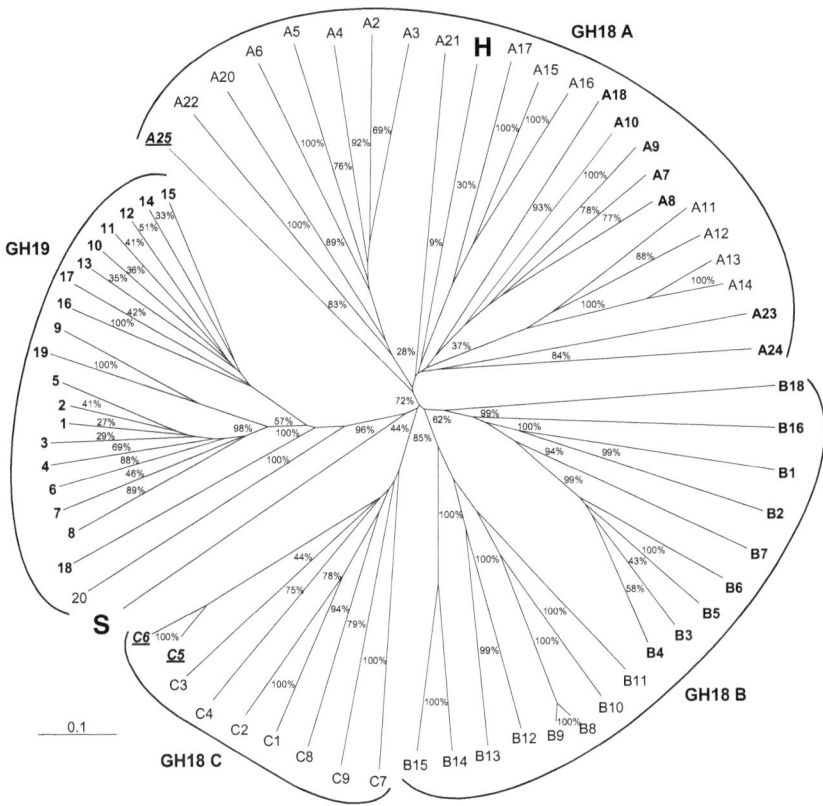

Figure 3.13: Amino acid sequence similarity tree based on 70 gene sequences from GH family 18 (n=50) and 19 (n=20). The phylogenetic tree was constructed using both the Neighbour-joining option of ClustalX (Thompson et al. 1997) as well as the Maximum-likelihood method of PROML (Phylip, version 3.6). Bootstrap values (1000 replicates) are given in *per cent* between the respective branches. Abbreviations: GH18 A: Subfamily 18A. GH18 B: Subfamily 18 B. GH18 C: Subfamily 18 C. GH19: GH family 19. **S**: *Sulfolobus tokodaii chi* gene sequence. **Eukaryal sequences:** Printed in bold. **_Euryarchaeal_** sequences: Printed in bold italics and underlined. **H**: *Halobacterium salinarum chi* gene sequence. Sequence abbreviations and respective Genbank accession numbers are given in the appendix (Tab. A3).

4. Discussion

This work focused on three main goals: The establishment of a comprehensive chitinase test panel, the screening of natural occurring bacterial communities and strain collections for chitinolytic active bacteria and the detection, isolation and characterisation of archaeal chitinases. These three goals are interconnected, with the chitinase test panel as focal point (Fig. 1.4).

4.1. The chitinase test panel

A novel approach of combined cultivation and molecular screening for chitinases was established in this work. The different parts of the test panel were used to elucidate archaeal chitinases. Furthermore, focus was given to cultivated bacteria with the genetic potential for chitinolysis and to the presence of excreted chitinases and their respective specific activity.

Until now cultivated chitinolytic strains were not tested concerning the presence of a genetic chitinase motive and strains that were screened genetically were not tested for their actual growth on chitin as sole carbon and nitrogen source (Cottrell et al. 1999, Cottrell & Kirchman 2000). Within this work the cultivation of microorganisms on chitin as sole carbon and nitrogen source was combined for the first time with molecular screening of the cultivated strains for a genetic chitinase motif. In addition, all strains that were able to grow under laboratory conditions with chitin as sole carbon and nitrogen source, were tested for the secretion of chitinases into the cultivation broth and the specific activity of the respective chitinases were determined. The molecular screening and evaluation of respective enzymes was also used for the detection and characterisation of archaeal chitinases.

Isolation of chitinolytic bacteria

For the establishment of the cultivation approach a homogenised *Palaemon adspersus* sample from the Baltic Sea as source for chitinolytic bacteria (see 2.1., Tab. 2.1) was used. The classic approach for the isolation of microorganisms is the

Discussion

application of solid media (Madigan et al. 2006). Hence, a solid minimal medium containing chitin was designed and established (see 3.1.). The solid chitin medium was most suitable for the initial isolation of bacteria and to obtain single colonies and subsequently pure cultures. However, the occurrence of clearing zones (see 3.1.) as a direct proof of chitinolytic activity did take up to 48 days and 58% of all growing strains did not show the formation of a clearing zone at all. The first clearing zones were detected after 15 days. This very long incubation time lead on the one hand to the drying of the isolation medium and on the other hand made fast decisions concerning the chitinolytic capabilities of the respective strains impossible. Hence, a second liquid minimal chitin medium with chitin as sole carbon and nitrogen source was developed and applied (see 3.1.) for the detection of chitinolytic bacteria.

It was shown that the solid chitin medium designed in this work was successfully used for the initial isolation of chitinolytic bacteria and to obtain pure cultures of chitinolytic strains. For isolation purposes in general a complex medium is used to account for the many occurring different nutritional needs of the bacterial strains (Madigan et al. 2006). Hence, this is a novel approach for the specific isolation of novel chitin degrading bacteria.

Detection of chitinolytic potential / activity

Within this second step of the screening panel molecular- and cultivation techniques were used. For the cultivation approach homogenised sediment samples from the eastern Mediterranean Deep Sea (see 2.1., Tab. 2.1) were used for the establishment and testing of the liquid chitin medium. After the initial isolation on chitin medium, the obtained pure strains were transferred into liquid chitin medium. Strains capable to grow in liquid chitin medium were regarded as chitinolytic.

The cultivation conditions under which a bacterial community is isolated can greatly influence the resulting cultivated bacterial strains (Eilers et al. 2000). Hence, it was tested if the subsequent cultivation in liquid chitin medium would change the bacterial composition by favouring a specific bacterial group. This was not the case.

Discussion

From the 36 isolated strains, 27 could be re-cultivated, corresponding to a recovery rate of 75 %. All phylogenetic classes present in the solid chitin medium were also present in the liquid chitin medium (see 3.1.). Only the number of *Gammaproteobacteria* differed and dropped from 20 strains growing on solid chitin medium to 11 strains growing in liquid chitin medium. However, the *Gammaproteobacteria* still dominated the isolated bacterial community (see Tab. 3.2 and Fig. 3.3).

The liquid chitin medium designed in this study was successfully used for the subsequent cultivation of chitinolytic pure cultures and gave direct evidence for the presence of chitinolysis as chitin was the only carbon and nitrogen source contained in the medium. The combination of solid and liquid cultivation techniques is a valid tool to obtain and validate chitin degrading bacteria from natural habitats and strain collections with regard to the test panel.

For the molecular detection of chitinolysis the glycoside hydrolase families were used. The glycoside hydrolase sub-family 18 A (GH18A) is regarded as the main chitinase family amongst bacteria (Metcalfe et al. 2002). Until now three sub-families of the GH18 family are known; GH18 A, B and C. Hobel and co-workers (2005) designed two degenerated primer sets according to the consensus motif of subfamily 18 A. These primer sets were used in this work to establish the molecular screening for the chitinolytic potential of bacterial isolates from natural habitats and strain collections (see 2.6. and 3.1.). To cover the natural occurring variations of the GH18A motif that is based in the possibility of the conserved exchange of amino acids and the resulting variations in the DNA sequence, as well as to compensate the natural variations occurring in the amino acid sequence of enzyme-parts not crucial for the correct function of the protein, two degenerated primer sets were used. The PCR protocols (see 2.6.) were successfully implemented and tested on the same 36 strains isolated from homogenised Mediterranean Deep Sea sediment sample, used for the testing of the liquid chitin medium (see 3.1.). It is noteworthy that 38.9 % of the tested strains were positive for the GH18A motif, but 52.8 % of the isolated strains were capable to

Discussion

grow under laboratory conditions on chitin as sole carbon and nitrogen source and did not possess the genetic GH18A motif. This finding is of great importance considering the composition of chitinolytic bacterial communities in natural habitats. As most studies are using only molecular methods to evaluate the chitinolytic bacterial community (LeCleir et al. 2004, Hobel et al. 2005, Lian et al. 2007, Bhuiyan et al. 2011) the findings of this study indicate that a greater proportion of chitinolytic bacteria can be expected in natural habitats.

With regard to the detection of archaeal chitinases, another approach was used, called genetic mining. For this approach not only the chitinase core motif, but a hole chitinase amino acid sequence was used to search the NCBI Genbank database. This search tactics enabled the detection of a novel crenarchaeal chitinase that would have not been found with a core motif search alone.

Biochemical proof of chitinolytic activity

Chitinases are a valuable tool to decrease the growing amount of waste produced by the shellfish production, thereby providing the basis for novel products, such as sweeteners, growth factors and single cell protein (Mejía-Saulés et al. 2006). Chitinases themselves can also be used in pest control such as insects, nematodes or moulds as reported by Mejia (2006) and references within. The used substrate, chitin itself, is literally too big to be digested by microorganisms internally. Thus enzymes have to be located at the outside of the microorganisms cell wall, or secreted into the environment as an extracellular enzyme. In addition to this reason, with respect to a possible fast and easy downstream processing of the detected chitinases, it was decided to lay special focus on the detection and evaluation of the activity of extracellular chitinases within the search for bacterial chitin degraders.

For this reason, the culture broth of bacteria cultivated in liquid chitin medium was investigated. After removal of cells and all non-soluble fractions the protein fraction of the culture broth was gained and the chitinase activity was determined. The chitinase activity was evaluated using the MBTH method (see 3.1.). The method was mo-

Discussion

dified to suit the needs of a rapid and protein saving approach. It was accomplished to miniaturise the procedure, allowing for the use of only one fifth of the amounts of chemicals and disposables as compared to the published method by Horn and co-workers (2004). The assay was scaled from 1 ml down to 200 µl end volume. By using double and triplicate measurements, the risk of errors due to the smaller volumes was overcome. The final volume of 200 µl enabled the use of a plate reader for the necessary photometrical measurement. Hence, not only the amount of necessary enzymes and chemicals was reduced, but also the time, as 96 samples were prepared at once and measured within a few minutes. Two of the 8 strains screened during the establishment of the assay were shown to excrete an extracellular chitinase into the cultivation medium (see Tab. 3.3). Whether this is a high or low number of chitinases excreting bacteria in the test sample can not be evaluated as no literature covering the amount of bacteria excreting chitinases exist until now. As the established method is very sensitive and designed to give specific activity values of excreted chitinases, it is very suitable to reveal bacterial strains with the potential for future easy chitinase production in a larger scale. The use of excreted chitinases in the biotechnological large scale production of chitinases has to the knowledge of the author not been used or reported yet.

The established MBTH assay was also used to evaluate the chitinolytic activities of the respective archaeal chitinases.

Decision criteria for chitinolytic active bacteria

After reviewing the collected data of the cultivation and genetic approach it can be clearly seen that only a combined decision criterion leads to a comprehensive picture of chitin degraders in a cultivatable bacterial community. Additionally, valid criteria are given for decision purposes. During the cultivation in liquid medium it was shown that the composition of the bacterial community considering the distribution among different phylogenetic classes remained the same. However the number of *Gammaproteobacteria* decreased. This could be attributed to the inability of the re-

Discussion

spective strains to utilise chitin as sole carbon and nitrogen source or to the inability of the strains to adapt to the liquid cultivation conditions. To strengthen the decision whether a bacterial strain is chitinolytic or not, another criterion is needed. This criterion is the presence of the GH18 A motif in the strains. Again, if only considering the presence of the genetic motif, 39.9% of the test-strains would qualify as potentially chitinolytic active.

In combination with the established detection and evaluation method of excreted chitinases the established screening panel can be used as a valuable tool to decide fast and reliably if a specific strain has not only chitinolytic potential but might even be of use in a future biotechnological process for the production of chitinases.

4.2. Screening of new environmental isolates and strains from collections

Using the established chitinase test panel, the chitinolytic activity and the respective genetic potential of natural occurring marine bacterial communities were elucidated and strains collections were screened for chitinolytic bacteria. In addition, all screened strains were tested for the presence of an excreted extracellular chitinase.

90.7 % (133 strains) of the strains possessed chitin degrading potential. This high number of chitinolytic bacteria was expected, as the sources for the strains were chosen for a maximum yield of chitin degrading strains. During the sampling in the different habitats, the strains were initially isolated on chitin minimal medium. *Actinomycetes* are known chitin degraders and the surface of bryozoans was supposed to favour chitinolytic strains due to the specific micro habitat, as discussed below. Considering further on only these positively tested strains it can be seen that 52 % (69 strains) possessed the GH18A motive, but only 29 % (38 strains) of these strains grew under laboratory conditions on chitin as sole carbon and nitrogen source, while 48 % (64 strains) utilised chitin as only carbon and nitrogen source while not possessing the GH18A motif (Fig. 4.1).

Discussion

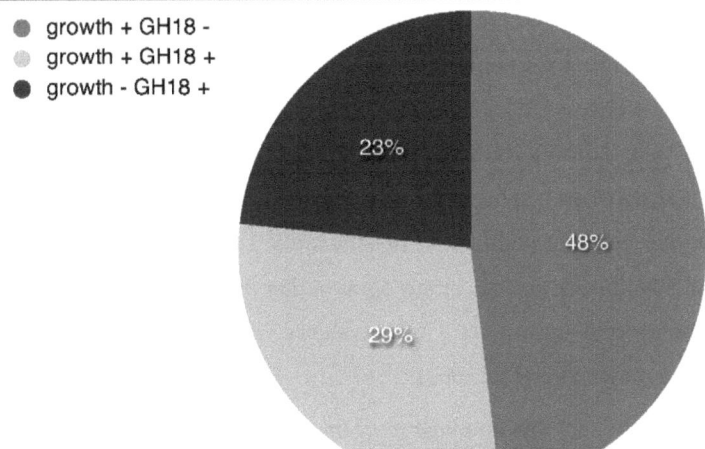

Figure 4.1: Chart depicting the distribution of potential bacterial chitin degraders from the marine environment based on 133 bacterial strains possessing a chitinolytic potential. Grey: Strains growing on chitin as sole carbon and nitrogen source, but not possessing the GH18A motif (n=64). Dark grey: Strains only possessing the GH18A motif, but not growing on chitin as sole carbon and nitrogen source (n=31). Light grey: Strains possessing the GH18A motif and growing on chitin as sole carbon and nitrogen source (n=38).

Discussion

This shows the clear domination of the GH18A motif within the chitin degrading bacteria. Even when assuming that the GH 19 family does not contribute at all to the tested strains and only GH 18 is considered, an even distribution of the strains amongst the three known GH18 classes A, B and C would lead to an expected maximum 33 % share of each sub-family. The detected 52 % of GH18A containing strains hence clearly show the domination of subfamily 18A among cultivatable bacterial strains and strengthens the prevailing opinion that the GH18 A subfamily dominates among chitin degrading bacteria. This prevalence was already proposed based on molecular studies (Metcalfe et al. 2002).

It is noteworthy that this domination of the GH18a motif would not have been detected with cultivation based methods alone. Only 29 % of the chitinolytic bacteria possessing the GH18A motif were capable to grow on chitin as sole carbon and nitrogen source, while 48 % of the strains growing on chitin as sole carbon and nitrogen source did not possess the GH18A motif. These results further strengthen the necessity of the established combined detection approach.

Comparison of marine habitats

The investigated marine habitats were eastern Mediterranean Deep Sea sediments (36 strains), western Mediterranean Deep Sea sediments (18 strains) and carapaces of the Baltic Sea shrimp *Palaemon adspersus* (14 strains). Due to the high amount of chitin produced annually in the oceans, it can be expected that chitin is abundant in the sediments and is used as nutritional source. Further, the carapaces of shrimp consist of chitin and are also expected to be used as nutritional source for microorganisms. Hence, it was decided to isolate bacteria from these sources on chitin medium and screen them for their chitin degrading capability.

Within all three habitats chitin degrading bacteria were found. They showed a distinct and unique distribution pattern of the chitin degrading strains amongst the phylogenetic classes. The eastern Mediterranean sediment and Baltic Sea shrimp sample were dominated by *Gammaproteobacteria* (eastern Mediterranean sediment: 51.1 %,

Discussion

Baltic Sea shrimp: 53.8 %), while the western Mediterranean sediment sample was dominated by *Actinobacteria* (38.5 %, see Fig. 3.5). The prevalence of *Gammaproteobacteria* in the eastern Mediterranean sediment sample is unexpected, as Gooday reported in his review (Gooday 1990a) a domination of actinomycetes among bacterial chitin degraders in sediment samples, as was detected in the western Mediterranean sediment sample. In general, a bacterial distribution with a major part of *Gammaproteobacteria* is typical for sediments (Hongxiang et al. 2008, Pachiadaki et al. 2010). The detected differences might be based on the different habitats. Although both sediment samples were taken from the deep sea, the western sediment sample originated from 600 m while the western sediment samples originated from up to 4400 m, with the most shallow samples taken in 2200 m. Hence, the organic materials that were sinking down to the sediment experienced different exposure times to the sea water. Hence, the microbial community settling on the different sedimenting particles is expected to differ, as was also reported by Ploug (2008) and references within. This in return would influence the bacterial community found in the respective sediments. In addition, the different sedimentation times would have also altered the composition of the degradable biological materials, thereby creating a natural depletion of different nutrients in the deep sea sediment and hence influences the bacterial composition capable of coping with the prevailing conditions. The differences in the bacterial communities from homogenised shrimp carapaces can be explained by the different microhabitat. By providing nutrients in the form of chitin from the carapace, and abundant aeration by the natural movement of the shrimps the development of a unique bacterial community can be favoured. A unique community of shrimp associated bacteria has been also described by Verschuere (2000) and references within.

The results of this study show a habitat specific distribution of chitin degrading bacteria at different habitats and strengthen the idea of specific chitin degrading communities in different habitats as proposed by LeCleir (2004).

Discussion

Screening of the KiWiZ strain collection

Bryozoan-derived bacteria (45 strains) and the *Actinomycetes* (31 strains) from the KiWiZ strain collection were screened for chitinolytic bacteria, as bryozoans are like decapods known chitin producers and may thereby also harbour a chitin degrading community, while *Actinomycetes* are already known to utilise chitin. The selected strains proved to be a good source for chitinolytic bacteria. Within the bryozoan-derived isolates 93.3 % of the screened strains possessed chitinolytic capabilities, while within the *Actinomycetes* 96.8 % of the screened strains possessed chitinolytic capabilities. In both cases, the phylogenetic distribution pattern of selected strains matched the pattern of chitinolytic strains (see Fig. 3.6 and Fig. 3.7). No shifts towards a specific phylogenetic class or genera were detected. This indicates that the chitinolytic bacterial community found on bryozoans is very versatile and the capability to degrade chitin is wide spread among the *Actinomycetes*.

The highly variable bacterial community from bryozoans that is influenced by the surrounding water (Pukall et al. 2001) might be the reason for the high number of chitinolytic bacteria within the bryozoan-derived isolates. In addition, bryozoans are known to use chitin as skeletal element (Taveners.R & Williams 1972). Thereby bryozoans would provide chitin as nutritional source and create a unique environment for a highly variable bacterial community. Hence, with the obtained results it was shown that bryozoans are a valuable source for chitin degrading bacteria.

The high number of chitinolytic bacteria within the *Actinomycetes* can be attributed to the isolation method of the strains and to the bacterial class itself. *Actinomycetes* are not only known producers of secondary metabolites but do also degrade chitin (Gomes et al. 2000). Hence, specific media for the isolation of *Actinomycetes* include the addition of chitin to enhance the yield (Hsu & Lockwood 1975). Such a medium was also used for the initial isolation of several of the strains in the KiWiZ strain collection. With 63.3 % most strains belonged to the genus *Streptomyces*. This prevalence of *Streptomycetes* within chitinolytic *Actinomycetes* was also reported by Ka-

Discussion

wase (2004) and might be attributed to the high multiplicity of the chitinase gene within the *Streptomycetes* species (Saito et al. 2003). The results of this study further strengthen the role of *Actinomycetes* and in special *Streptomycetes* strains as source for the discovery of chitinolytic bacteria.

Excreted chitinases

As already mentioned above (see 1.2. and Fig 1.2) chitinases initiate the chitinolytic process within bacteria and fungi. Chitin itself is to big to be transported as polymer into the bacterial cell. Hence, excreted chitinases can be of great importance for the initial colonisation of a surface, not just allowing the use of chitin as nutritional source, but also hindering fungi to colonise the same surface.

Of the 102 strains capable to grow in liquid chitin medium, 11.8 % excreted a detectable chitinase into the cultivation broth (see Tab 3.4). Most of the strains capable to excrete a detectable chitinase belonged to the class *Actinomycetes* and within this class to the genus *Streptomyces*. These results strengthen the role of the *Actinomycetes* as a novel source for chitinases. The highest specific activity was measured in the chitinase from the strain *Streptomyces* sp. HB122 with 0.42 mU/mg. This is in the lower range of chitinase activities when comparing it to the activity of other actinomycetes chitinases such as the chitinase from *Streptomyces thermoviolaceus*, possessing a thermostable chitinase. This strain showed chitinase activities of 4000 mU/mg of the crude, non purified enzyme. A maximum of 8250 mU/mg was reached with the purified enzyme (Tsujibo et al. 1993).

The established detection method can not only be used for biotechnological purposes but also to elucidate the possible ecological role of excreted chitinases in the initial colonisation of surfaces and their role as defence against fungal competitors.

Phylogenetic analysis of potential chitin degrading isolates

To go into greater detail of the phylogenetic relationships of the chitin degrading cultivated bacteria used in this study, a phylogenetic analysis was conducted, comparing the 16S rDNA. For this reason high quality sequences are necessary. In total 95

Discussion

(Tab. A2, Appendix) sequences were obtained from strains isolated during the different sampling campaigns in the Baltic Sea, the Mediterranean and the bryozoan-derived isolates of the KiWiZ strain collection. Also strains isolated from bryozoans were included to give a broader overview of cultivated bacterial chitinolytic strains. The respective bacterial sequences were compared with the closest related type strains within a phylogenetic tree (Fig. 3.9A-C). The five phylogentic clusters (see 3.2.) showed the specific distribution of chitin degrading bacteria according to origin as proposed by (LeCleir et al. 2004) and also the presence of chitinolytic bacterial species in all habitats. The specific distribution can be seen in the region and host specific clusters 1 with isolates from only Baltic Sea bryozoans (Fig. 3.9A) and in the geographical region specific cluster 2 (Fig. 3.9A), containing only Mediterranean strains but from different sampling sites (bryozoans and deep sea sediments). Three species specific clusters (Fig. 3.9C), cluster 3 containing only *Pseudoalteromonas* species, cluster 4 containing only *Pseudomonas* species and cluster 5 containing only *Bacillus* species, show the broad distribution of chitinolytic bacterial species in all habitats, that are geographically separated from each other.

These results show that the identified chitin degrading bacteria build specifically associated communities in certain habitats. Additionally, omnipresent chitinolytic bacterial strains exist.

Discussion

4.3. Archaeal chitinases

Using molecular screening and biochemical proof of chitinolytic activity of enzymes established within the test panel in this work, the first crenarchaeal chitinase from *Sulfolobus tokodaii* was detected, overexpressed purified and initially characterised and the euryarchaeal chitinase from *Halobacterium salinarum* was overexpressed, purified and initially characterised. The latter cloning and expression was done for a chitinase for the first time in a bacterial mesohaline expression system by maintaining halophilic activity. The activity of the recombinant *Halobacterium salinarum* chitinase was demonstrated with the established MBTH method.

Within the domain of archaea, only ten euryarchaeal chitinases were found so far in terms of genetic or molecular information (Henrissat 1991). Most of them were annotated by genome mining in analogy to known genes. Thus, their actual chitin degrading capabilities have not been elucidated yet. To our knowledge, enzymatic characterisation including proof of activity has been done only for four organisms. The activities of the *Sulfolobus tokodaii* (75 mU/mg) and *Halobacterium salinarum* (38 mU/mg) proteins were in the same range as the activities of the reported euryarchaeal chitinases of the *Pyrococcus furiosus* and *Thermococcus kodakaraensis* KOD1. The *Pyrococcus furiosus* chitinases ChiA and ChiB showed a specific activity of 35 mU/mg (Gao et al. 2003), whereas the *Thermococcus kodakaraensis* KOD1 chitinase had a specific activity of 18 mU/mg (Tanaka et al. 1999). The chitinolytic activities of the detected *Sulfolobus tokodaii* and *Halobacterium salinarum* chitinases were forty to eighty-fold lower as compared to the natively purified chitinase from *Thermococcus chitinophagus* with a specific activity of 3 U/mg. This chitinase is the only non-recombinant archaeal chitinase described so far (Andronopoulou & Vorgias 2004). The difference of the detected chitinolytic activity may be inherent to the technique of recombinant protein production. It might be overcome by using the archaeal expression systems with e.g. *Sulfolobus solfataricus* as host (Albers et al. 2006) instead of *E. coli*. In addition, a chitinase was described in the genome of *Halobacterium salinarum*

Discussion

strain NRC-1 (Hatori et al. 2006). The enzyme was reported to be halophilic, with an optimal activity at about 1M NaCl. The activity was retained at salt concentrations ranging up to approximately 5M NaCl. The enzyme was insensitive to DMSO concentrations of up to 30% (v/v). However, a specific activity was not given and hence could not be compared to our results.

Discussion

4.3.1. *Sulfolobus tokodaii*

The *Sulfolobus tokodaii chi* sequence

Chitinase related sequence motives were found in the ORF BAB65950 of *Sulfolobus tokodaii*. However, the GH18 motif could not be clearly detected in the sequence of *Sulfolobus tokodaii*. Hence, also the GH19 motif was aligned to the *Sulfolobus tokodaii* sequence. The glycoside hydrolase families 18 and 19 show completely different structures and are not related. Neither the GH18 specific TIM barrel structure, nor the bilobal structure of the GH19 family with its high alpha helical content could be detected in the predicted secondary structure (see 3.3.). Also, the affiliation of the protein to the chitobiase family (GH20) is unlikely, as the GH20 enzyme class contains also the TIM barrel structure within its catalytic domain (Tews et al. 1996) and does not act primarily on polymeric substrates (Hoell et al. 2010).

The detected BTB-domain (broad-complex, tramtrack and bric a brac) is known as a protein-protein interaction motif (Bardwell & Treisman 1994) and plays a role in dimerisation. According to this, the identified BTB domain is proposed as linking region. The adjacent chitin/cellulose binding domain (ChtBD3) is known from many different glycoside hydrolase enzymes (Brun et al. 1997) and might be crucial for the enzyme's carboactive properties. The detected highly conserved fibronectin type III domain is typical for bacterial chitinases and reported to be also involved in substrate binding (Toratani et al. 2006). It remains open, whether the enzyme is excreted or not, as no leader sequence was detected.

Properties of the *Sulfolobus tokodaii* chitinase

The native molecular weight of the *Sulfolobus tokodaii* protein was determined to be 130 kDa indicating a homodimeric protein structure. This is in good accordance with the detected BTB domain. A dimeric status is also known from other chitinases, such as the *Pyrococcus furiosus* chitinase (Nakamura et al. 2007).

Discussion

The *Sulfolobus tokodaii* chitinase showed optimal activity at 70 °C and pH 2.5 and hence was classified as thermoacidophilic. This enzyme optimum corresponds to the natural living conditions of *Sulfolobus tokodaii* dwelling in sulfur rich hot acid springs in volcanic regions. Its optimal growth conditions are aerobic, at 80 °C with a low pH (2-3) (Suzuki et al. 2002).

Discussion

4.3.2. *Halobacterium salinarum*

Although the chitinolytic properties of the *chi* gene of *Halobacterium salinarum* have been published by Hatori (2006) and co-workers shortly after the respective sequence was retrieved in this work, this is the first study expressing the enzyme in a bacterial mesohaline expression system while maintaining halophilic activity.

The *Halobacterium salinarum chi* sequence

The *Halobacterium salinarum* chi sequence of the respective ORF CAP 13543 could be clearly attributed to the glycoside hydrolase family 18, with the highly conserved DXDXE motif located at the C-terminal end. The chitin/cellulose binding domain (ChtBD3) was also detected in the *Halobacterium salinarum* sequence, including the conserved binding residues Trp_{35} and Tyr_{41}. The found polycystic kidney diseases I domain (PKD) was proposed to function as a ligand binding site in protein-protein or protein-carbohydrate interactions, which was also found in bacterial chitinases (Orikoshi et al. 2005). No leader sequences were detected, hence the question remains whether the protein is excreted or not.

Properties of the *Halobacterium salinarum* chitinase

The native molecular weight of the *Halobacterium salinarum* chitinase was determined to be 330 kDa, indicating a hexameric protein. This unusual conformation may be an experimental artefact, since many halophilic proteins have unusual elution properties due to their highly charged surfaces (Oren et al. 2000).

A specific activity of the purified protein of 38 mU/mg was detected at pH 6 and 28 °C with 0.02 M $MgCl_2$ and 10% (v/v) glycerol. A need for a high osmotic pressure environment was demonstrated, which allows for the proper folding of the resulting enzyme (Ramalingam et al. 1992). These findings correspond well with the living conditions of *Halobacterium salinarum* in highly saline environments (Colwell et al. 1979).

4.3.3. Comparison of chitinases sequences

While the amino acid sequence of the *Halobacterium salinarum* chitinase could be clearly attributed to the glycoside hydrolase family 18 A, the amino acid sequence of the *Sulfolobus tokodaii* chitinase was different from the glycoside hydrolase families 18 and 19 (Fig. 3.13). This was supported by the phylogenetic analysis. Concerning sequence homologies, molecular and kinetic properties, euryarchaeal chitinases were highly similar to their bacterial and eukaryal counterparts. Hence, euryarchaeal chitinases formed no domain specific group as described for other archaeal enzymes of sugar metabolism, such as glucokinases or pyruvate kinases (Siebers & Schönheit 2005). The crenarchaeal sequence of *Sulfolobus tokodaii* did neither cluster clearly with the GH18 nor the GH19 family. As this is the first crenarchaeal sequence available yet, more sequences have to be investigated and included in the protein tree. However, the detected *Sulfolobus tokodaii* chitinase might enable the finding of more chitinases from the domain of crenarchaea.

5. Future perspectives

Within this work a combined approach for the elucidation of chitinolytic microorganisms was established and evaluated. With more sampling campaigns and higher sample numbers the arisen question of habitat specific bacterial communities will be answered. The successful miniaturisation and automation of the MBTH test for the chitinase activity paves the way for future high throughput screening assays and can be utilised in the rapid and fast evaluation of chitinases from various sources and applications. The purification and description of the identified extracellular chitinases is one of the next goals. The respective bacterial strains might even hold the potential for a rapid industrial production of chitinases. Considering archaeal chitinases, the characterisation of the two found chitinases is not complete yet. Especially the *Sulfolobus tokodaii* chitinase might bring new insights into the chitinolytic processes found in crenarchaeota, maybe this is the first evidence of a novel chitinolytic mechanism, unique to crenarchaeotes. This work is just the starting point for many experiments and further analysis of chitinases from different and even extreme environments.

6. Bibliography

Albers SV, Jonuscheit M, Dinkelaker S, Urich T and others (2006) Production of recombinant and tagged proteins in the hyperthermophilic archaeon *Sulfolobus solfataricus*. Applied and Environmental Microbiology 72:102-111

Altschul SF, Gish W, Miller W, Myers EW, Lipman DJ (1990) Basic local alignment search tool. Journal of Molecular Biology 215:403-410

Andronopoulou E, Vorgias CE (2004) Isolation, cloning, and overexpression of a chitinase gene fragment from the hyperthermophilic archaeon *Thermococcus chitonophagus*: Semi-denaturing purification of the recombinant peptide and investigation of its relation with other chitinases. Protein Expression and Purification 35:264-271

Anthon GE, Barrett DM (2002) Determination of reducing sugars with 3-methyl-2-benzothiazolinonehydrazone. Analytical Biochemistry 305:287-289

Bardwell VJ, Treisman R (1994) The poz domain: A conserved protein-protein interaction motif. Genes & Development 8:1664-1677

Bhuiyan FA, Nagata S, Ohnishi K (2011) Novel chitinase genes from metagenomic DNA prepared from marine sediments in southwest Japan. Pakistan Journal of Biological Science 14:204-211

Blaak H, Schnellmann JR, Walter S, Henrissat B, Schrempf H (1993) Characterization of an exochitinase from *Streptomyces olivaceoviridis*, its corresponding gene, putative protein domains and relationship to other chitinases. European Journal of Biochemistry / FEBS 214:659-669

Blackwell J (1988) Physical methods for the determination of chitin structure and conformation. Methods in Enzymology 161:435-442

Boulanger A, Dejean G, Lautier M, Glories M, Zischek C, Arlat M, Lauber E (2010) Identification and regulation of the N-Acetylglucosamine utilization pathway of the plant pathogenic bacterium *Xanthomonas campestris* pv. *Campestris*. Journal of Bacteriology 192:1487-1497

Bibliography

Bradford MM (1976) A rapid and sensitive method for the quantitation of microgram quantities of protein utilizing the principle of protein-dye binding. Analytical Biochemistry 72:248-254

Bräuer S, Cadillo-Quiroz H, Ward RJ, Yavitt J, Zinder S (2010) *Methanoregula boonei* gen. nov., sp. nov., an acidiphilic methanogen isolated from an acidic peat bog. International Journal of Systematic and Evolutionary Microbiology

Brun E, Moriaud F, Gans P, Blackledge MJ, Barras F, Marion D (1997) Solution structure of the cellulose-binding domain of the endoglucanase Z secreted by *Erwinia Chrysanthemi*. Journal of Biochemistry 36:16074-16086

Bussink AP, van Eijk M, Renkema GH, Aerts JM, Boot RG (2006) The biology of the gaucher cell: The cradle of human chitinases. International Reviews in Cytology 252:71-76

Cantarel BL, Coutinho PM, Rancurel C, Bernard T, Lombard V, Henrissat B (2009) The carbohydrate-active enzymes database (CAZy): An expert resource for glycogenomics. Nucleic Acids Research 37:233-238

Colwell RR, Litchfield CD, Vreeland RH, Kiefer LA, Gibbons NE (1979) Taxonomic studies of red halophilic bacteria. International Journal of Systematic Bacteriology 29:379-399

Cottrell MT, Kirchman DL (2000) Community composition of marine bacterioplankton determined by 16s rRNA gene clone libraries and fluorescence in situ hybridization. Applied and Environmental Microbiology 66:5116-5122

Cottrell MT, Moore JA, Kirchman DL (1999) Chitinases from uncultured marine microorganisms. Applied and Environmental Microbiology 65:2553-2553

Ehrlich H, Simon P, Carrillo-Cabrera W, Bazhenov VV and others (2010) Insights into chemistry of biological materials: Newly discovered silica-aragonite-chitin biocomposites in demosponges. Journal of Chemical Materials 22:1462-1471

Eijsink VGH, Vaaje-Kolstad G, Westereng B, Horn SJ, Liu ZL, Zhai H, Sorlie M (2010) An oxidative enzyme boosting the enzymatic conversion of recalcitrant polysaccharides. Science 330:219-222

Eilers H, Pernthaler J, Amann R (2000) Succession of pelagic marine bacteria during enrichment: A close look at cultivation-induced shifts. Applied and Environmental Microbiology 66:4634-4640

Elkins JG, Podar M, Graham DE, Makarova KS and others (2008) A korarchaeal genome reveals insights into the evolution of the archaea. Proceedings of the National Academy of Sciences 105:8102-8107

Funkhouser JD, Aronson NN (2007) Chitinase family GH18: Evolutionary insights from the genomic history of a diverse protein family. BMC Evolutionary Biology 7:96-96

Gao J, Bauer MW, Shockley KR, Pysz MA, Kelly RM (2003) Growth of hyperthermophilic archaeon *Pyrococcus furiosus* on chitin involves two family 18 chitinases. Applied and Environmental Microbiology 69:3119-3128

Garcia E, Johnston D, Whitaker JR, Shoemaker SP (1993) Assessment of endo-1,4-beta-d-glucanase activity by a rapid colorimetric assay using disodium 2,2'-bicinchoninate. Journal of Food Biochemistry 17:135-145

Gärtner A, Blümel M, Wiese J, Imhoff JF (2011) Isolation and characterisation of bacteria from the eastern Mediterranean Deep Sea. Antonie Van Leeuwenhoek 100:421-435

Gasteiger E, Alexandre G, Hoogland C, Ivanyi I, Apple RD, Bairoch A (2003) Expasy: The proteomics server for in-depth protein knowledge and analysis. Nucleic Acids Research 31:3784-3788

Gomes RC, Semedo LT, Soares RM, Alviano CS, Linhares LF, Coelho RR (2000) Chitinolytic activity of actinomycetes from a cerrado soil and their potential in biocontrol. Letters in Applied Microbiology 30:146-150

Gooday GW (1990a) The ecology of chitin degradation. Advances in Microbial Ecology 11:387-430

Gooday GW (1990b) Physiology of microbial degradation of chitin and chitosan. Biodegradation 1:177-190

Guindon S, Gascuel O (2003) A simple, fast, and accurate algorithm to estimate large phylogenies by maximum likelihood. Systematic Biology 52:696-704

Guindon S, Lethiec F, Duroux P, Gascuel O (2005) PhyML online - a web server for fast maximum likelihood-based phylogenetic inference. Nucleic Acids Research 33:557-559

Harrison FC, Kennedy ME (1922) The red discoloration of cured cod fish. Proceedings and Transactions of the Royal Society of Canada 3:101-152

Hatori Y, Sato M, Orishimo K, Yatsunami R, Endo K, Fukui T, Nakamura S (2006) Characterization of recombinant family 18 chitinase from extremely halophilic archaeon *Halobacterium salinarum* strain NRC-1. Chitin and Chitosan Research 12:201-201

Heindl H, Wiese J, Thiel V, Imhoff JF (2010) Phylogenetic diversity and antimicrobial activities of bryozoan-associated bacteria isolated from Mediterranean and Baltic Sea habitats. Systematic Applied Microbiology 33:94-104

Henrissat B (1991) A classification of glycosyl hydrolases based on amino acid sequence similarities. Biochemical Journal 280:309-316

Henrissat B, Davies G (1997) Structural and sequence-based classification of glycoside hydrolases. Current Opinion in Structural Biology 7:637-644

Hobel CFV, Marteinsson VT, Hreggvidsson GO, Kristjansson JK (2005) Investigation of the microbial ecology of intertidal hot springs by using diversity analysis of 16s rRNA and chitinase genes. Applied and Environmental Microbiology 71:2771-2776

Hoell IA, Vaaje-Kolstad G, Eijsink VG (2010) Structure and function of enzymes acting on chitin and chitosan. Biotechnology and Genetic Engineering Reviews 27:331-366

Hongxiang X, Min W, Xiaogu W, Junyi Y, Chunsheng W (2008) Bacterial diversity in Deep-Sea sediment from northeastern Pacific Ocean. Acta Ecologica Sinica 28:479-485

Hood MA (1991) Comparison of four methods for measuring chitinase activity and the application of the 4-muf assay in aquatic environments. Journal of Microbiological Methods 13:151-160

Horn SJ, Eijsink VGH (2004) A reliable reducing end assay for chito-oligosaccharides. Carbohydrate Polymers 56:35-39

Hsu SC, Lockwood JL (1975) Powdered chitin agar as a selective medium for enumeration of actinomycetes in water and soil. Applied and Environmental Microbiology 29:422-422

Huber R, Stohr J, Hohenhaus S, Rachel R, Burggraf S, Jannasch HW, Stetter KO (1995) *Thermococcus chitonophagus* sp. nov., a novel, chitin-degrading, hyperthermophillic archaeum from a Deep-Sea hydrothermal vent environment. Archives of Microbiology 164:255-264

Huet Jl, Rucktooa P, Clantin B, Azarkan M, Looze Y, Villeret V, Wintjens R (2008) X-ray structure of papaya chitinase reveals the substrate binding mode of glycosyl hydrolase family 19 chitinases. Biochemistry 47:8283-8291

Imoto T, Yagishita K (1971) A simple activity measurement of lysozyme. Agricultural and Biological Chemistry 35:1154-1156

Jolles P, Muzzarelli RAA (eds) (1999) Chitin and Chitinases, Birkhäuser, Basel/Switzerland

Bibliography

Karlsson M, Stenlid J (2008) Comparative evolutionary histories of the fungal chitinase gene family reveal non-random size expansions and contractions due to adaptive natural selection. Evolutionary Bioinformatics Online 4:47-60

Karlsson M, Stenlid J (2009) Evolution of family 18 glycoside hydrolases: Diversity, domain structures and phylogenetic relationships. Journal of Molecular Microbiology and Biotechnology 16:208-223

Kasprzewska A (2003) Plant chitinases - regulation and function. Cell Molecular Biology Letters 8:809-824

Kawarabayasi Y (2001) Complete genome sequence of an aerobic thermoacidophilic crenarchaeon, *Sulfolobus tokodaii* strain 7. DNA Research 8:123-140

Kawase T, Saito A, Sato T, Kanai R and others (2004) Distribution and phylogenetic analysis of family 19 chitinases in actinobacteria. Applied and Environmental Microbiology 70:1135-1144

Keyhani NO, Roseman S (1999) Physiological aspects of chitin catabolism in marine bacteria. BBA-General Subjects 1473:108-122

Keyhani NO, Wang LX, Lee YC, Roseman S (2000) The chitin disaccharide, n,n'-diacetylchitobiose, is catabolized by *Escherichia coli* and is transported/phosphorylated by the Phosphoenolpyruvate:Glucose phosphotransferase system. Journal of Biological Chemistry 275:33084-33090

Kezuka Y, Ohishi M, Itoh Y, Watanabe J, Mitsutomi M, Watanabe T, Nonaka T (2006) Structural studies of a two-domain chitinase from *Streptomyces griseus* HUT 6037. Journal of Molecular Biology 358:472-484

Kirchman DL, White J (1999) Hydrolysis and mineralization of chitin in the Delaware estuary. Aquatic Microbial Ecology 18:187-196

Kublanov IV, Perevalova AA, Slobodkina GB, Lebedinsky AV and others (2008) Biodiversity of thermophilic prokaryotes with hydrolytic activities in hot springs of Uzon caldera, Kamchatka (Russia). Applied and Environmental Microbiology 75:286-291

Lämmli UK (1970) Cleavage of structural proteins during the assembly of the head of bacteriophage T4. Nature 227:680-685

Lane D (1991) 16s/23s rRNA sequencing. In: Stackebrabd E, Goodfellow M (eds) Techniques in bacterial systematics. John Wiley & Sons, New York, p 115-175

LeCleir GR, Buchan A, Hollibaugh JT (2004) Chitinase gene sequences retrieved from diverse aquatic habitats reveal environment-specific distributions. Applied and Environmental Microbiology 70:6977-6983

Lian M, Lin S, Zeng R (2007) Chitinase gene diversity at a Deep Sea station of the east Pacific nodule province. FEBS Journal 11:463-467

Lloyd AJ, Brandish PE, Gilbey AM, Bugg TDH (2004) Phospho-n-acetyl-muramyl-pentapeptide translocase from *Escherichia coli*: Catalytic role of conserved aspartic acid residues. The Journal of Bacteriology 186:1747-1757

Ludwig W, Strunk O, Westram R, Richter Land others (2004) ARB: A software environment for sequence data. Nucleic Acids Research 32:1363-1371

Madigan MT, Martinko JM, Brock TD (2006) Brock biology of microorganisms, Pearson Prentice Hall, Upper Saddle River, NJ

McGuffin LJ, Bryson K, Jones DT (2000) The psipred protein structure prediction server. Bioinformatics 16:404-405

Mejía-Saulés JE, Waliszewski KN, Garcia MA, Cruz-Camarillo R (2006) The use of crude shrimp shell powder for chitinase production by *Serratia marcescens* wf. Food Technology and Biotechnology 44:95-100

Metcalfe AC, Krsek M, Gooday GW, Prosser JI, Wellington EM (2002) Molecular analysis of a bacterial chitinolytic community in an upland pasture. Applied and Environmental Microbiology 68:5042-5050

Miller JH (1972) Experiments in molecular genetics. Cold Spring Harbor Laboratory, New York, p 352-355

Muyzer G, Dewaal EC, Uitterlinden AG (1993) Profiling of complex microbial populations by denaturing gradient gel electrophoresis analysis of polymerase chain reaction-amplified genes coding for 16s rRNA. Applied and Environmental Microbiolog 59:695-700

Nakamura T, Mine S, Hagihara Y, Ishikawa K, Uegaki K (2007) Structure of the catalytic domain of the hyperthermophilic chitinase from *Pyrococcus furiosus*. Acta Crystallographica Section F, Structural Biology and Crystallization Communications 63:7-11

Ng WV, Kennedy SP, Mahairas GG, Berquist B and others (2000) Genome sequence of *Halobacterium* species NRC-1. Proceedings of the National Academy of Sciences of the United States of America 97:12176-12181

O'Neil MJ (2006) The Merck index : An encyclopedia of chemicals, drugs, and biologicals, Merck, Whitehouse Station, NJ

Oku T, Ishikawa K (2006) Analysis of the hyperthermophilic chitinase from *Pyrococcus furiosus*: Activity toward crystalline chitin. Bioscience, Biotechnology, and Biochemistry 70:1696-1701

Orikoshi H, Nakayama S, Hanato C, Miyamoto K, Tsujibo H (2005) Role of the n-terminal polycystic kidney disease domain in chitin degradation by chitinase a from a marine bacterium, *Alteromonas* sp. Strain o-7. Journal of Applied Microbiology 99:551-557

Pachiadaki MG, Lykousis V, Stefanou EG, Kormas KA (2010) Prokaryotic community structure and diversity in the sediments of an active submarine mud volcano (Kazan mud volcano, east Mediterranean Sea). FEMS Microbiology Ecology 72:429-444

Park JK, Keyhani NO, Roseman S (2000) Chitin catabolism in the marine bacterium *Vibrio furnissii* - identification, molecular cloning, and characterization of a n-n'-diacetylchitobiose phosphorylase. Journal of Biological Chemistry 275:33077-33083

Patil RS, Ghormade V, Deshpande MV (2000) Chitinolytic enzymes: An exploration. Enzyme and Microbial Technology 26:473-483

Pfeiffer F, Schuster SC, Broicher A, Falb M and others (2008) Evolution in the laboratory: The genome of *Halobacterium salinarum* strain R1 compared to that of strain NRC-1. Genomics 91:335-346

Ploug H, Iversen MH, Fischer G (2008) Ballast, sinking velocity, and apparent diffusivity within marine snow and zooplankton fecal pellets: Implications for substrate turnover by attached bacteria. Limnology and Oceanography 53:1878-1886

Poulicek M, Gaill Fo, Goffinet G (1998) Chitin biodegradation in marine environments. In: Stankiewicz BA, van Bergen PF (eds), Vol 707. American Chemical Society, Washington, DC, p 163-210

Prakash UNA, Jayanthi M, Sabarinathan R, Kangueane P, Mathew L, Sekar K (2010) Evolution, homology conservation, and identification of unique sequence signatures in GH19 family chitinases. Journal of Molecular Evolution 70:466-478

Pukall R, Kramer I, Rohde M, Stackebrandt E (2001) Microbial diversity of cultivatable bacteria associated with the North Sea bryozoan *Flustra foliacea*. Systematic and Applied Microbiology 24:623-633

Ramaiah N, Hill RT, Chun J, Ravel J, Matte MH, Straube WL, Colwell RR (2000) Use of a *chiA* probe for detection of chitinase genes in bacteria from the Chesapeake Bay. FEMS Microbiology Ecology 34:63-71

Ramalingam K, Aimoto S, Bello J (1992) Conformational studies of anionic melittin analogues: Effect of peptide concentration, pH, ionic strength, and temperature-models for protein folding and halophilic proteins. Biopolymers 32:981-992

Bibliography

Ruiz-Herrera J, Martinez-Espinoza AD (1999) Chitin biosynthesis and structural organization in vivo. In: Chitin and Chitinases. Jolles, P, Muzzarelli, RAA. (eds) Basel, Switzerland: Birkhäuser, p 39-53

Saito A, Fujii T, Miyashita K (2003) Distribution and evolution of chitinase genes in *Streptomyces* species: Involvement of gene-duplication and domain-deletion. Antonie Van Leeuwenhoek 84:7-15

Salzer P, Bonanomi a, Beyer K, Vögeli-Lange R and others (2000) Differential expression of eight chitinase genes in *Medicago truncatula* roots during mycorrhiza formation, nodulation, and pathogen infection. Molecular Plant-Microbe Interactions 13:763-777

Sambrook J, Fritsch EF, Maniatis T, Russell RW (2001) Molecular cloning: A laboratory manual, Cold Spring Harbor Laboratory Press, Cold Spring Harbor

Sanger F, Nicklen S, Coulson AR (1977) DNA sequencing with chain-terminating inhibitors. Proceedings of the Natural Academic Society of the USA 74:5463-5467

Saunders E, Tindall BJ, Fähnrich R, Lapidus A and others (2010) Complete genome sequence of *Haloterrigena turkmenica* type strain (4KT). Standards in Genomic Sciences 2:107-116

Scopes RK (1994) Protein purification : Principles and practice, Springer-Verlag, New York

Shuhui L, Mok YK, Wong WSF (2009) Role of mammalian chitinases in asthma. International Archives of Allergy and Immunology 149:369-377

Siebers B, Schönheit P (2005) Unusual pathways and enzymes of central carbohydrate metabolism in archaea. Current Opinion in Microbiology 8:695-705

Suzuki K, Taiyoji M, Sugawara N, Nikaidou N, Henrissat B, Watanabe T (1999) The third chitinase gene (chic) of *Serratia marcescens* 2170 and the relationship of its product to other bacterial chitinases. Biochemical Journal 343:587-596

Suzuki T, Iwasaki T, Uzawa T, Hara K and others (2002) *Sulfolobus tokodaii* sp. nov. (f. *Sulfolobus* sp. strain 7), a new member of the genus *Sulfolobus* isolated from Beppu hot springs, Japan. Extremophiles 6:39-44

Tanaka T, Fujiwara S, Nishikori S, Fukui T, Takagi M, Imanaka T (1999) A unique chitinase with dual active sites and triple substrate binding sites from the hyperthermophilic archaeon *Pyrococcus kodakaraensis* KOD1. Applied and Environmental Microbiology 65:5338-5344

Tanaka T, Fukui T, Atomi H, Imanaka T (2003) Characterization of an exo-[beta]-d-glucosaminidase involved in a novel chitinolytic pathway from the hyperthermophilic archaeon *Thermococcus kodakaraensis* KOD1. Journal of Bacteriology 185:5175-5181

Tanaka T, Fukui T, Fujiwara S, Atomi H, Imanaka T (2004) Concerted action of diacetylchitobiose deacetylase and exo-beta-d-glucosaminidase in a novel chitinolytic pathway in the hyperthermophilic archaeon *Thermococcus kodakaraensis* KOD1. The Journal of Biological Chemistry 279:30021-30027

Tanaka T, Fukui T, Imanaka T (2001) Different cleavage specificities of the dual catalytic domains in chitinase from the hyperthermophilic archaeon *Thermococcus kodakaraensis* KOD1. The Journal of Biological Chemistry 276:35629-35635

Taveners.R, Williams A (1972) Secretion and structure of skeleton of living and fossil bryozoa. Philosophical Transactions of the British Society Biological Sciences 264:97-160

Tews I, Perrakis A, Oppenheim A, Dauter Z, Wilson KS, Vorgias CE (1996) Bacterial chitobiase structure provides insight into catalytic mechanism and the basis of Tay-Sachs disease. Natural and Structural Biology 3:638-648

Thiel V, Neulinger SC, Staufenberger T, Schmaljohann R, Imhoff JF (2007) Spatial distribution of sponge-associated bacteria in the mediterranean sponge *Tethya aurantium*. FEMS Microbiology Ecology 59:47-63

Bibliography

Thompson JD, Gibson TJ, Plewniak F, Jeanmougin F, Higgins DG (1997) The clustal_x windows interface: Flexible strategies for multiple sequence alignment aided by quality analysis tools. Nucleic Acids Research 25:4876-4882

Tindall BJ, Schneider S, Lapidus A, Copeland A and others (2009) Complete genome sequence of *Halomicrobium mukohataei* type strain (arg-2t). Standards in Genomic Sciences 1:270-277

Toratani T, Kezuka Y, Nonaka T, Hiragi Y, Watanabe T (2006) Structure of full-length bacterial chitinase containing two fibronectin type III domains revealed by small angle x-ray scattering. Biochemical and Biophysical Research Communications 348:814-818

Tsuji H, Nishimura S, Inui T, Kado Y, Ishikawa K, Nakamura T, Uegaki K (2010) Kinetic and crystallographic analyses of the catalytic domain of chitinase from *Pyrococcus furiosus*- the role of conserved residues in the active site. The FEBS Journal 277:2683-2695

Tsujibo H, Kubota T, Yamamoto M, Miyamoto K, Inamori Y (2003) Characterization of chitinase genes from an alkaliphilic actinomycete, *Nocardiopsis prasina* OPC-131. Applied and Environmental Microbiology 69:894-900

Tsujibo H, Minoura K, Miyamoto K, Endo H, Moriwaki M, Inamori Y (1993) Purification and properties of a thermostable chitinase from *Streptomyces thermoviolaceus* OPC-520. Applied and Environmental Microbiology 59:620-622

Verschuere L, Rombaut G, Sorgeloos P, Verstraete W (2000) Probiotic bacteria as biological control agents in aquaculture. Microbiology and Molecular Biology Reviews 64:655-671

Wierenga RK (2001) The TIM-barrel fold: A versatile framework for efficient enzymes. FEBS Letters 492:193-198

Xiao X, Yin XB, Lin H, Sun LG, You ZY, Wang P, Wang FP (2005) Chitinase genes in lake sediments of Ardley island, Antarctica. Applied and Environmental Microbiology 71:7904-7909

Zor T, Selinger Z (1996) Linearization of the Bradford protein assay increases its sensitivity: Theoretical and experimental studies. Proteins 308:302-308

7. Appendix

Table A1: List of strains used during this study. B: Baltic Sea; M: Mediterranean Sea; W: Western; E: Eastern; S: KiWiZ strain collection; vis. deg.: Visible degradation of chitin; "+": positive; "-": negative.

ID	Strain Name	Origin	Spec. activity [mU/mg]	ChiA1 primer	ChiA2 primer	vis. deg.
AB145	Arthrobacter sp.	M S	0	+	-	-
B150	Cellulomonas sp. BCHID458428	B S	0	-	-	-
B157	Shewanella sp. LMG 23025	B S	0.12	-	-	-
B163	Shewanella sp. F15	B S	0	-	-	-
B167	Marinomonas sp. NJ522	B S	0	+	-	-
B171	Shewanella sp. F15	B S	0	-	-	-
B206	Alteromonadaceae bacterium E1	B S	0	-	-	-
B22	Pseudoalteromonas sp. EH-2-1	B S	0	-	-	-
B278	Alteromonas macleodii 'Deep ecotype'	M S	0	-	-	-
B304	Pseudoalteromonas sp. ws17	M S	0	-	-	-
B345	Alteromonas macleodii DSM 6062	M S	0	-	-	-
B350	Sphingomonas sp. SKJH-30	M S	0	+	-	-
B390	Tenacibaculum adriaticum B390T	M S	0	-	-	-
B411	Pseudovibrio ascidiaceicola F423(= NBRC 100514)	M S	0	-	-	-
BB1	Sphingopyxis sp. ME-BiC05060	B S	0	+	-	-
BB11a	Streptomyces sp. 3490	B S	0.25	+	-	+
BB13	Gamma proteobacterium UMB21A	B S	0	+	-	-

Appendix

ID	Strain Name	Origin	Spec. activity [mU/mg]	ChiA1 primer	ChiA2 primer	vis. deg.
BB15	*Marinobacter* sp. ASs2019	B S	0	-	-	-
BB17	*Erythrobacter aquimaris* strain SW-110	B S	0	-	-	-
BB18	*Amorphus coralli* strain RS.Sph.026	B S	0	+	-	-
BB2	*Roseobacter* sp. P123	B S	0	+	-	-
BB21	Bacterium DG1026	B S	0	+	-	-
BB22	*Roseovarius aestuarii* SMK-122	B S	0	+	-	-
BB23	*Jannaschia pohangensis* haplotype H1-M8	B S	0	-	-	-
BB25	*Microbacterium* sp. MOLA 56	B S	0	+	-	-
BB32	*Erythrobacter* sp. CNU001	B S	0	-	-	-
BB35	Glacial ice bacterium SB12K-2-16	B S	0	-	-	-
BB37	Glacial ice bacterium SB12K-2-16	B S	0	+	-	-
BB4	*Sphingopyxis* sp. ME-BiC05060	B S	0	+	-	-
BB40	Rhodobacteraceae bacterium P92	B S	0	+	-	-
BB41	*Bacillus hwajinpoensis* SW-72	B S	0	+	-	-
BB44	*Microbulbifer thermotolerans* JAMM 1340	B S	0	-	-	-
BB46	*Sphingopyxis* sp. ME-BiC05060	B S	0	+	-	-
BB50 a	*Streptomyces* sp. 060386	B S	*0.04*	-	-	+
BB51 b	*Paracoccus* sp. HZ04	B S	0	-	-	-
BB52	*Bacillus licheniformis* strain HNL09	B S	0	-	-	-
BB57	*Mycobacterium sacrum* BN 3151	B S	0	-	-	-

Appendix

ID	Strain Name	Origin	Spec. activity [mU/mg]	ChiA1 primer	ChiA2 primer	vis. deg.
BB63	*Pseudonocardia* sp. BHF008	B S	0	-	-	-
BB66 b	*Halomonas* sp. NT N13	B S	0	+	-	-
BB72 a	*Arthrobacter parietis* LMG 22281	B S	0	+	-	-
BB75 b	*Exiguobacterium* sp. ARCTIC-P28	B S	0	-	-	-
BB77	*Arthrobacter* sp. VTT E-052904	B S	0	-	-	-
BB79	*Pseudomonas stutzeri* ZoBell ATCC 14405	B S	0	+	-	-
BB80	*Pseudomonas stutzeri* ZoBell ATCC 14405	B S	0	+	-	-
BB82 b	*Halomonas* sp. TNB I18	B S	0	-	-	-
BB85	*Halomonas* sp. NT N95	B S	0	-	-	-
D02	*Stenotrophomonas maltophilia*	M E	0	+	+	-
D04	*Bacillus circulans*	M E	0	-	-	-
D06	*Marinobacter flavimaris*	M E	0	-	-	-
D112	*Halobacillus karajiensis*	M E	0	-	-	-
D113	*Marinobacter salsuginis*	M E	0	+	+	-
D114	*Pseudoalteromonas tetraodonis* strain Do-17	M E	0	+	-	-
D117	*Leeuwenhoekiella blandensis*	M E	0	-	-	-
D33	*Alteromonas litorea*	M E	0	-	-	-
D34	*Pseudoalteromonas elyakovii*	M E	U	+	I	-
D35a	*Bacillus subtilis*	M E	0	-	-	-
D35x	*Marinobacter flavimaris*	M E	0	-	+	-

Appendix

ID	Strain Name	Origin	Spec. activity [mU/mg]	ChiA1 primer	ChiA2 primer	vis. deg.
D36	*Marinobacter flavimaris*	M E	0	-	-	-
D37	*Pseudomonas stutzeri*	M E	0	-	-	-
D38	*Pseudomonas stutzeri*	M E	0	-	-	-
D39	*Alteromonas macleodii*	M E	0	+	-	-
D42	*Alteromonas marina* strain SW-47	M E	0	+	+	-
D43	*Pseudomonas stutzeri*, strain 4C68	M E	0	-	+	-
D44	*Micrococcus luteus*	M E	0	-	-	-
D45	*Alteromonas addita*	M E	0	-	-	-
D47	*Alteromonas addita*	M E	0	-	-	-
D48	*Alteromonas marina* strain SW-47	M E	0	-	-	-
D49	*Stenotrophomonas maltophilia*	M E	0	+	+	-
D50	*Pseudomonas stutzeri*	M E	0	-	-	-
D52	*Bacillus axarquiensis*	M E	0	-	-	-
D56	*Alteromonas addita*	M E	0	-	-	-
D81	*Streptomyces sampsonii*	M E	0.03	-	+	+
D81a	*Bacillus arsenicus*	M E	0	-	-	-
D92	*Streptomyces sampsonii*	M E	0	-	-	-
HB096	*Streptomyces* sp. AY079156	B S	0	-	-	+
HB100	*Streptomyces* sp. AF112174	B S	0	-	-	-
HB107	*Streptomyces platensis*	B S	0	+	+	-

Appendix

ID	Strain Name	Origin	Spec. activity [mU/mg]	ChiA1 primer	ChiA2 primer	vis. deg.
HB117	*Streptomyces* sp. AF112174	B S	0	-	+	+
HB122	*Streptomyces* sp. VTT E-99-1/ *S. griseus*	B S	*0.42*	+	-	+
HB130	*Streptomyces* sp. AF112174	B S	0	-	-	+
HB132	*Streptomyces lavendulae*	B S	0	-	-	+
HB138	*Streptomyces lavendulae*	B S	0	+	+	+
HB141	*Nocardiopsis alba* (X97883)	B S	0	-	-	+
HB147	unidentified bacterium	B S	0	-	-	+
HB149	*Streptomyces* sp. AF112174	B S	0	-	-	+
HB157	*Streptomyces* sp. AY079156	B S	0	+	-	+
HB180	*Streptomyces collinus*	B S	0	-	-	+
HB181	*Streptomyces sanglieri*	B S	0	-	-	+
HB200	*Streptomyces microstreptospora*	B S	0	-	-	+
HB202	*Streptomyces* sp. VTT E-99-1326 (A4)	B S	0	+	-	+
HB213	*Pseudomonas* sp. AY209180	B S	0	-	-	-
HB217	unidentified bacterium	B S	*0.04*	+	+	+
HB225	*Streptomyces* sp. U93338	B S	0	+	+	+
HB239	*Micromonospora* sp.	B S	0	-	-	-
HB241	*Micromonospora* sp.	B S	0	-	-	+
HB243	*Streptomyces microstreptospora*	B S	0	-	+	-
HB244	*Bacillus pumilus* AB098578	B S	0	-	-	-

Appendix

ID	Strain Name	Origin	Spec. activity [mU/mg]	ChiA1 primer	ChiA2 primer	vis. deg.
HB254	*Micromonospore* sp.	B S	0	-	-	-
HB298	*Streptomyces* sp. AJ391831	B S	0	-	-	+
HB318	*Streptomyces peuceticus*	B S	0	+	+	-
HB346	*Streptomyces roseoflavus*	B S	0.11	-	-	+
HB372	*Lechevalieria fradiae*	B S	0	+	-	+
HB374	*Lechevalieria fradiae*	B S	0		+	-
HB375	*Micromonospora* sp.	B S	0	+	+	+
HB383	*Prauseria* sp. TUT1202	B S	0	-	-	+
i34a	*Pseudomonas chloritidismutans*	M W	0	-	-	-
i34b	*Pseudomonas chloritidismutans*	M W	0	-	-	-
i35a	*Flavobacterium* sp.	M W	0	-	-	-
i35b	*Marinobacter flavimaris* strain SW-145	M W	0	-	-	-
i36	*Rhodococcus cercidiphyllus* strain YIM 65003	M W	0	-	-	-
i37	*Pseudomonas moraviensis* strain CCM 7280	M W	0	+	-	-
i59	*Kocuria palustris*	M W	0	-	-	-
i60	*Paracoccus yeeii* strain G1212	M W	0	+	-	-
i61	*Gordonia terrae* (T); ATCC 25594T	M W	0	-	-	-
i62	*Micrococcus indicus*, type strain BBQ1	M W	0.10	-	-	-
i69	*Corynebacterium terpenotabidum*	M W	0	+	+	-
i70	*Corynebacterium variabilis* strain DSM 20132	M W	0	-	-	-

Appendix

ID	Strain Name	Origin	Spec. activity [mU/mg]	ChiA1 primer	ChiA2 primer	vis. deg.
i71	Corynebacterium terpenotabidum	M W	0	-	-	-
i77	Bacillus hwajinpoensis	M W	0	-	-	-
i82	Rhodococcus yunnanensis	M W	0	-	-	-
i83	Pseudomonas chloritidismutans	M W	0	-	-	-
i84	Erythrobacter litoralis HTCC2594	M W	0	-	-	-
i85	Aurantimonas coralicida strain WP1	M W	0	-	-	-
PAD1	Exiguobacterium sp.	B	0.01	-	-	-
PAD10	Pseudoalteromonas sp.	B	0	-	-	-
PAD11	Pseudoalteromonas sp.	B	0	-	-	-
PAD12	Pseudoalteromonas sp.	B	0	+	-	-
PAD16	Vibrio sp.	B	0.39	-	-	-
PAD20	Pseudoalteromonas sp.	B	0	-	+	-
PAD27	Pseudoalteromonas sp.	B	0.12	-	-	-
PAD28	Pseudoalteromonas sp.	B	0	+	+	-
PAD29	Ralstonia detusculanense	B	0	-	+	+
PAD30	Pseudoalteromonas citrea	B	0	-	-	+
PAD33	Micrococcus luteus	B	0	-	-	-
PAD34	Aquimarina muelleri	B	0	+	+	-
PAD36	Micrococcus luteus	B	0	-	-	-
PAD37	Cellulophaga pacifica	B	0	-	-	-

Appendix

ID	Strain Name	Origin	Spec. activity [mU/mg]	ChiA1 primer	ChiA2 primer	vis. deg.
S06	Streptomyces flavofuscus strain NRRL B-8036	M W	0	+	+	-
S08	Arthrobacter tecti	M W	0	-	-	-
S086	Arthrobacter tecti	M W	0	-	-	-
S09	Pseudoalteromonas elyakovii strain BSi20610	M W	0	-	-	+
S10	Bacillus novalis	M W	0	-	-	-
S11	Bacillus foraminis	M W	0.02	-	-	-
S33	Bacillus decolorationis	M W	0	-	-	-
S33x	Bacillus firmus	M W	0	-	-	-

Appendix

Table A2: List of strains used for the phylogenetic analysis

ID	Closest type strain	Acc. Nr.	Similarity	Origin
B150	*Cellulosimicrobium funkei* strain W6122	AY501364	99.73%	Baltic Sea
B157	*Shewanella baltica* NCTC10735	AJ000214	97.35%	Baltic Sea
B163	*Shewanella denitrificans* strain OS-217	AJ311964	97.15%	Baltic Sea
B167	*Marinomonas pontica* strain 46-16	AY539835	97.95%	Baltic Sea
B171	*Shewanella denitrificans* strain OS-217	AJ311964	98.49%	Baltic Sea
B206	*Shewanella frigidimarina* ACAM 591T	U85903	97.76%	Baltic Sea
B22	*Pseudoalteromonas elyakovii*	AF082562	99.93%	Baltic Sea
B278	*Alteromonas litorea*	AY428573	98.07%	Mediterranean
B304	*Pseudoalteromonas aurantia*	X82135	97.41%	Mediterranean
B345	*Alteromonas marina* strain SW-47	AF529060	98.03%	Mediterranean
B350	*Sphingomonas pseudosanguinis* type strain G1-2T	AM412238	98.67%	Mediterranean
B390	*Tenacibaculum adriaticum* type strain B390	AM412314	100%	Mediterranean
B411	*Pseudovibrio japonicus*	AB246748	99.07%	Mediterranean
BB1	*Sphingopyxis flavimaris* strain SW-151	AY554010	97.32%	Baltic Sea
BB11a	*Streptomyces rubrogriseus* strain: NBRC 15455	AB184681	99.17%	Baltic Sea
BB13	*Psychrobacter piscatorii*	AB453700	99.83%	Baltic Sea
BB18	*Amorphus coralli* strain RS.Sph.026	DQ097300	95.84%	Baltic Sea
BB2	*Ruegeria atlantica*	D88526	97.39%	Baltic Sea
BB21	*Pelagibius litoralis* strain CL-UU02	DQ401091	93.16%	Baltic Sea
BB22	*Roseovarius aestuarii* strain SMK-122	EU156066	97.56%	Baltic Sea
BB23	*Jannaschia rubra* type strain 4SM3T	AJ748747	97.13%	Baltic Sea

Appendix

ID	Closest type strain	Acc. Nr.	Similarity	Origin
BB25	Microbacterium deminutum	AB234026	98.91%	Baltic Sea
BB32	Erythrobacter longus strain DSM 6997	AF465835	97.22%	Baltic Sea
BB37	Mycobacterium aurum strain ATCC 23366	FJ172298	99.19%	Baltic Sea
BB4	Sphingopyxis flavimaris strain SW-151	AY554010	97.39%	Baltic Sea
BB40	Ruegeria scottomollicae type strain LMG 24367T	AM905330	98.47%	Baltic Sea
BB41	Bacillus hwajinpoensis	AF541966	98.86%	Baltic Sea
BB44	Microbulbifer thermotolerans strain: JAMB A94.	AB124836	99.55%	Baltic Sea
BB46	Sphingopyxis flavimaris strain SW-151	AY554010	97.18%	Baltic Sea
BB50a	Streptomyces griseorubens strain: NBRC 12780.	AB184139	99.81%	Baltic Sea
BB51b	Paracoccus homiensis strain DD-R11	DQ342239	97.57%	Baltic Sea
BB52	Bacillus licheniformis ATCC 14580	CP000002	99.86%	Baltic Sea
BB57	Mycobacterium frederiksbergense strain DSM 44346.	AJ276274	99.37%	Baltic Sea
BB63	Pseudonocardia carboxydivorans strain Y8	EF114314	99.72%	Baltic Sea
BB66b	Halomonas neptunia strain Eplume1	AF212202	98.38%	Baltic Sea
BB72a	Arthrobacter tumbae, type strain LMG 19501	AJ315069	98.81%	Baltic Sea
BB75b	Exiguobacterium oxidotolerans	AB105164	99.48%	Baltic Sea
BB77	Arthrobacter agilis	X80748	98.52%	Baltic Sea
BB79	Pseudomonas stutzeri	AF094748	99.75%	Baltic Sea
BB80	Pseudomonas stutzeri	AF094748	99.82%	Baltic Sea
BB82b	Halomonas boliviensis strain LC1	AY245449	99.21%	Baltic Sea
BB85	Halomonas neptunia strain Eplume1	AF212202	98.56%	Baltic Sea
D02	Stenotrophomonas maltophilia, DSM50170T (AY484506)	FM992709	99.4%	Mediterranean Deep Sea

Appendix

ID	Closest type strain	Acc. Nr.	Similarity	Origin
D04	Bacillus circulans, DSM11T (AY724690)	FM992802	97%	Mediterranean Deep Sea
D112	Marinobacter salsuginis strain SD-14B	AY881246	99%	Mediterranean Deep Sea
D113	Pseudoalteromonas tetraodonis strain Do-17	EF028328	97%	Mediterranean Deep Sea
D114	Alteromonas litorea, JCM12188T (AY428573)	AB257325	100%	Mediterranean Deep Sea
D117	Leeuwenhoekiella blandensis	DQ294291	99%	Mediterranean Deep Sea
D33	Alteromonas litorea, JCM12188T (AY428573)	FM992780	97.4%	Mediterranean Deep Sea
D34	Pseudoalteromonas elyakovii, KMM162T (AF082562)	FM992789	99.5%	Mediterranean Deep Sea
D35a(= D35)	Bacillus subtilis, DSM10T (AJ276351)	FM992801	99.9%	Mediterranean Deep Sea
D35x	Marinobacter flavimaris, DSM16070T (AY517632)	FM992844	98.9%	Mediterranean Deep Sea
D38	Pseudomonas stutzeri, ATCC17588T (AF094748)	FM992716	99.7%	Mediterranean Deep Sea
D39	Alteromonas marina, JCM11804T (AF529060)	FM992717	98.6%	Mediterranean Deep Sea
D42	Alteromonas marina, JCM11804T (AF529060)	FM992781	97.7%	Mediterranean Deep Sea
D43	Pseudomonas stutzeri, ATCC17588T (AF094748)	FM992782	100%	Mediterranean Deep Sea
D44	Micrococcus luteus, ATCC4698T (AF542073)	FM992718	99.3%	Mediterranean Deep Sea
D45	Alteromonas addita, LMG22532T (AY682202)	FM992719	97.2%	Mediterranean Deep Sea
D47	Alteromonas addita, LMG22532T (AY682202)	FM992720	96.5%	Mediterranean Deep Sea
D48	Alteromonas marina, JCM11804T (AF529060)	FM992783	98.8%	Mediterranean Deep Sea
D49	Stenotrophomonas maltophilia, DSM50170T (AY484506)	FM992721	99.4%	Mediterranean Deep Sea
D50	Pseudomonas stutzeri, ATCC17588T (AF094748)	FM992722	99.8%	Mediterranean Deep Sea
D52	Bacillus muralis, LMG20238T (AJ628748)	FM992833	99.5%	Mediterranean Deep Sea
D56	Alteromonas addita, LMG22532T (AY682202)	FM992724	97.3%	Mediterranean Deep Sea

Appendix

ID	Closest type strain	Acc. Nr.	Similarity	Origin
D81	*Streptomyces sampsonii*, ATCC25495T (D63871)	FM992731	99.9%	Mediterranean Deep Sea
D81a	*Bacillus barbaricus*, DSM14730T (AJ422145)	FM992811	99.6%	Mediterranean Deep Sea
D92	*Streptomyces sampsonii*, ATCC25495T (D63871)	FM992735	98.2%	Mediterranean Deep Sea
i34b	*Pseudomonas chloritidismutans*	AY017341	99%	Mediterranean Deep Sea
i36	*Rhodococcus cercidiphyllus* strain YIM 65003	EU325542	99%	Mediterranean Deep Sea
i37	*Pseudomonas moraviensis* strain CCM 7280	AY970952	99%	Mediterranean Deep Sea
i60	*Paracoccus yeeii* strain G1212	AY014173	99%	Mediterranean Deep Sea
i61	*Gordonia terrae* (T); ATCC 25594T	X81922	99%	Mediterranean Deep Sea
i69	*Micrococcus indicus*, type strain BBQ1	AM158920	99%	Mediterranean Deep Sea
i70	*Corynebacterium variabilis* (strain DSM 20132)	AJ222815	99%	Mediterranean Deep Sea
i77	*Bacillus hwajinpoensis*	AF541966	99%	Mediterranean Deep Sea
i83	*Pseudomonas chloritidismutans*	AY017341	99%	Mediterranean Deep Sea
i84	*Erythrobacter litoralis* HTCC2594	CP000157	97%	Mediterranean Deep Sea
i85	*Aurantimonas coralicida* strain WP1	AY065627	98%	Mediterranean Deep Sea
PAD1	*Exiguobacterium oxidotolerans*	AB105164	99.91%	Baltic Sea
PAD11	*Pseudoalteromonas haloplanktis*	X67024	99.31%	Baltic Sea
PAD16	*Vibrio calviensis*	AF118021	98.11%	Baltic Sea
PAD20	*Pseudoalteromonas paragorgicola*	AY040229	99.5%	Baltic Sea
PAD27	*Pseudoalteromonas haloplanktis*	X67024	97.02%	Baltic Sea
PAD28	*Pseudoalteromonas paragorgicola*	AY040229	96.53%	Baltic Sea
PAD33	*Micrococcus luteus*	AJ536198	99.2%	Baltic Sea

Appendix

ID	Closest type strain	Acc. Nr.	Similarity	Origin
PAD34	*Aquimarina muelleri*	AY608406	97.8%	Baltic Sea
PAD36	*Micrococcus luteus*	AJ536198	99.64%	Baltic Sea
PAD37	*Cellulophaga pacifica*	AB100840	99.82%	Baltic Sea
S06	*Streptomyces flavofuscus*, ATCC19908T (DQ026648)	FM992747	97.8%	Mediterranean Deep Sea
S08	*Arthrobacter tecti*, LMG22282T (AJ639829)	FM992748	98.8%	Mediterranean Deep Sea
S09	*Pseudoalteromonas elyakovii*, KMM162T (AF082562)	FM992775	99.5%	Mediterranean Deep Sea
S10	*Bacillus novalis*, DSM15603T (AJ542512)	FM992819	96.4%	Mediterranean Deep Sea
S11	*Bacillus foraminis*, LMG23174T (AJ717382)	FM992820	98.4%	Mediterranean Deep Sea
S33	*Bacillus decolorationis*, LMG19507T (AJ315075)	FM992824	97.8%	Mediterranean Deep Sea
S33x	*Bacillus infantis*, JCM13438T (AY904032)	FM992838	99.7%	Mediterranean Deep Sea

Appendix

Table A3: List of protein sequences used for phylogenetic analysis of GH 18 and GH19 Families.

Abb.	Sequence accession number	Organism
A2	gi|42782677|ref|NP_979924.1|	Bacillus cereus
A3	gi|51891675|ref|YP_074366.1|	Symbiobacterium thermophilum
A4	gi|51893120|ref|YP_075811.1|	Symbiobacterium thermophilum
A5	gi|106886714|ref|ZP_01354047.	Clostridium phytofermentans
A6	gi|52785710|ref|YP_091539.1|	Bacillus licheniformis
A7	Agi|145613938|ref|XP_363321.2|	Magnaporthe oryzae
A8	gi|39973265|ref|XP_368023.1|	Magnaporthe oryzae
A9	gi|42544812|gb|EAA67655.1|	Gibberella zeae PH-1
A10	gi|71020355|ref|XP_760408.1|	Ustilago maydis
A11	gi|71559190|gb|AAZ38189.1|	Agrotis segetum nucleopolyhedrovirus
A12	Gi: 29726689	Serratia marcescens
A13	gi|90577214|ref|ZP_01233025.1	Vibrio angustum
A14	gi|27359169|gb|AAO08114.1|	Vibrio vulnificus
A15	gi|83644004|ref|YP_432439.1|	Hahella chejuensis
A16	gi|89073061|ref|ZP_01159608.1	Photobacterium sp.
A17	gi|15641086|ref|NP_230718.1|	Vibrio cholerae O1 biovar El Tor str.
A18	gi|50308105|ref|XP_454053.1|	Kluyveromyces lactis
A20	gi|34498191|ref|NP_902406.1|	Chromobacterium violaceum
A21	gi|94970550|ref|YP_592598.1|	Candidatus Koribacter versatilis

Appendix

Abb.	Sequence accession number	Organism
A22	gi|67933425|ref|ZP_00526542.1	Solibacter usitatus
A23	gi|22328814|ref|NP_193707.2|	Arabidopsis thaliana
A24	gi|7291098|gb|AAF46534.1|	Drosophila melanogaster
A25	Gi: 154000271	Methanoregula boonei gen. nov., sp. nov.
B1	gi|39943038|ref|XP_361056.1|	Magnaporthe grisea
B2	gi|67902508|ref|XP_681510.1|	Aspergillus nidulans
B3	gi|10954033|gb|AAG25709.1|AF3	Malus x domestica
B4	gi|75708015|gb|ABA26457.1|	Citrullus lanatus
B5	gi|118200080|emb|CAJ43737.1|	Coffea arabica
B6	gi|2342435|dbj|BAA21861.1|	Arabidopsis thaliana
B7	gi|73622089|sp|Q53NL5.1|XIP2_	Oryza sativa
B8	gi|89072995|ref|ZP_01159542.1	Photobacterium sp.
B9	gi|90578969|ref|ZP_01234779.1	Vibrio angustum
B10	gi|90021349|ref|YP_527176.1|	Saccharophagus degradans
B11	gi|34498390|ref|NP_902605.1|	Chromobacterium violaceum
B12	gi|42782803|ref|NP_980050.1|	Bacillus cereus
B13	gi|83648108|ref|YP_436543.1|	Hahella chejuensis
B14	gi|66048001|ref|YP_237842.1|	Pseudomonas syringae pv. syringae
B15	gi|46908115|ref|YP_014504.1|	Listeria monocytogenes
B16	gi|9971103|emb|CAC07216.1|	Metarhizium acridum

Appendix

Abb.	Sequence accession number	Organism
B18	gi|50304909|ref|XP_452410.1|	*Kluyveromyces lactis*
C1	gi|100125808|ref|ZP_01331104.	*Burkholderia pseudomallei*
C2	gi|34496895|ref|NP_901110.1|	*Chromobacterium violaceum*
C3	gi|75759513|ref|ZP_00739603.1	*Bacillus thuringiensis* serovar israelensis
C4	gi|54298112|ref|YP_124481.1|	*Legionella pneumophila* str. Paris
C5	Gi: 126030289	*Pyrococcus furiosus.*
C6	Gi: 57641700	*Thermococcus kodakarensis* KOD1
C7	gi|89339030|ref|ZP_01191795.1	*Mycobacterium flavescens*
C8	gi|68231397|ref|ZP_00570566.1	*Frankia* sp.
C9	gi|89361969|ref|ZP_01199781.1	*Xanthobacter autotrophicus*
1	gi|1310915|pdb|2BAA|	*Hordeum vulgare*
2	Gi: 195927481	*Carica papaya*
3	gi|7435355|pir||T03032	*Zea mays*
4	gi|1705811|sp|P16579|CHI6_POPT	*Populus trichocarpa*
5	gi|4960049|gb|AAD34596.1|AF147	*Humulus lupulus*
6	gi|2108350|gb|AAC49718.1|	*Pinus strobus*
7	gi|6002766|gb|AAF00131.1|AF147	*Fragaria x ananassa*
8	gi|7595841|gb|AAF64475.1|AF241	*Cucumis melo*
9	gi|30691147|gb|AAO17294.1|	*Ficus carica*
10	gi|461740|sp|P80052|CHIT_DIOJA	*Dioscorea japonica*

Appendix

Abb.	Sequence accession number	Organism
11	gi|7488930|pir||T14341	*Daucus carota*
12	gi|116324|sp|P27054|CHI4_PHAVU	*Phaseolus vulgaris*
13	gi|1168935|sp|P42820|CHIP_BETV	*Beta vulgaris*
14	gi|4741848|gb|AAD28733.1|AF112	*Triticum aestivum*
15	gi|2570162|dbj|BAA22966.1|	*Chenopodium amaranticolor*
16	gi|7435352|pir||T09131	*Picea glauca*
17	gi|2129790|pir||S65778	*Brassica napus*
18	gi|23394444|gb|AAN31509.1|	*Phytophthora infestans*
19	gi|113510|sp|P11218|AGI_URTDI	*Urtica dioica*
20	gi|16759224|ref|NP_454841.1|	*Salmonella enterica*

i want morebooks!

Buy your books fast and straightforward online - at one of world's fastest growing online book stores! Environmentally sound due to Print-on-Demand technologies.

Buy your books online at
www.get-morebooks.com

Kaufen Sie Ihre Bücher schnell und unkompliziert online – auf einer der am schnellsten wachsenden Buchhandelsplattformen weltweit! Dank Print-On-Demand umwelt- und ressourcenschonend produziert.

Bücher schneller online kaufen
www.morebooks.de

 VDM Verlagsservicegesellschaft mbH
Heinrich-Böcking-Str. 6-8   Telefon: +49 681 3720 174   info@vdm-vsg.de
D - 66121 Saarbrücken   Telefax: +49 681 3720 1749   www.vdm-vsg.de

Printed by Books on Demand GmbH, Norderstedt / Germany